Animal Design

by John Sparks

With contributions by
John Napier and Martin Wells

BRITISH BROADCASTING CORPORATION

© The authors and the British Broadcasting Corporation 1972
First published 1972

Published by the British Broadcasting Corporation
35 Marylebone High Street, London WIM 4AA

Acknowledgment is due to WEIDENFELD AND NICOLSON LIMITED
for the chapter 'On being a parasite' from *Lower animals* by Dr Martin Wells

Printed in England by Westerham Press, Ltd, Westerham, Kent
ISBN: 0 563 10684 0

Contents

Published in conjunction with a series of ten television programmes *Animal Design*, first broadcast on BBC–1 on Thursdays at 11.15 pm starting 19 October 1972.

The programmes written and produced by John Sparks.

The illustrations and diagrams in this book are by Richard Orr.

Acknowledgement is due to the following for permission to reproduce photographs:

HEATHER ANGEL anemones, pages 9 and 10, sea slug, page 11, limpets, page 12; ARDEA vulture (P. Blasdale), page 86, owl (Werner Curth), page 89; DR C. K. BRAIN baboon, page 78; BRITISH MUSEUM (NATURAL HISTORY) backbone, page 32, skin model, page 62, dolphin models, page 63; CAMERA PRESS pony (Ray Hamilton), page 43; DR DAVID CHIVERS monkey, page 76; BRUCE COLEMAN LTD puffin (David & Katie Urry), front cover, squid (Jane Burton), page 12, dragon fish (Russ Kinne), page 35, zebras (Simon Trevor), pages 48–49, elephant (Simon Trevor), page 53, herd of elephants (C. A. W. Guggisberg), page 54, dolphins (Russ Kinne), page 71, stork (Peter Jackson), page 90, albatross (Francisco Erize), page 91, pelicans (Jane Burton), page 92; PETER HILL dolphins, page 72, starfish, back cover; M. J. D. HIRONS flea, louse, page 23; FRANK LANE swift (François Merlet), pheasant (Ronald Austing), page 93, humming-birds (Ronald Austing), page 94; NATIONAL INSTITUTE OF OCEANOGRAPHY baleen, page 70; NATIONAL RESEARCH COUNCIL OF CANADA anemones, from Canadian Journal of Zoology, 45, pp. 895–906 (1967), page 6; NATURAL SCIENCE PHOTOS spider (K. H. Hyatt), page 15; SPORT AND GENERAL horse racing, page 45; JOHN TOPHAM Suffolk horses, page 43; DAVID AND KATIE URRY gannet, page 87, puffins, page 88; DR D. P. WILSON tunny, page 37, pollack, bass, page 38; JOHN J. YEATS X-ray, page 51.

Introduction

All animals can be regarded as machines and must be as well designed as aeroplanes or motor-cars. The concept of 'design' is not meant to imply in this book that living things are the product of some heavenly inspired draughtsman; on the contrary. All but a few religious fanatics nowadays accept the notion that life has evolved slowly by a process of natural selection.

However, in order to survive and leave adequate off-spring, an animal's structure must be capable of carrying loads normally imposed on it throughout its life, because an animal is never exempt from the physical 'laws' of nature. The idea that the body's form is arrived at by a random process of trial and error does not alter the fact that at this moment in time animals can be recognised and assessed in engineering terms and 'design' seems to be the best word and, I hope, will be accepted without humbug.

The purpose of this book is to give those who have a general interest in natural history a broad insight into the mechanical problems faced by animals, and how these have been solved. The selection of subjects has been dictated chiefly by the material available for the television series and is accordingly rather biased towards the higher vertebrates.

Nevertheless, I have tried to give some idea of progression from the beginning to the end by starting with soft-bodied animals that live in water, and concluding with birds which perhaps represent the ultimate achievement in animal design.

By no stretch of the imagination is the text supposed to be the last word on animal mechanics. Authors, far more competent in bio-engineering than myself, have written several excellent books treating the subject in much greater depth and these are referred to in the bibliography. I am especially grateful to Dr Martin Wells and Dr John Napier for contributing their authoritative chapters on *Parasites* and *Man* respectively.

John Sparks
BBC Natural History Unit, Bristol

Behaviour of the anemone Actinostola *transferring from a tile to the shell of a mussel (*Modiolus*). The whole sequence took about 30 minutes.*

1

Strength without skeletons

It is not difficult to persuade someone who is dying from thirst in a hot, arid desert, with only mirages of oases for comfort, that life as we know it is utterly dependent upon water. Our bodies are composed more of the stuff than anything else. We do not however depend upon fluid for support, because, like so many animals we have a good internal chassis which will preserve our shape, and adequately bear the stresses and strains of everyday life. Not all animals are reinforced by skeletons of one form or another. Sea anemones, worms and some molluscs have soft, pliable bodies, which derive their power and strength by muscles squeezing on fluid. Some, which have strengthened bodies, even use a hydraulic method of shunting liquid around to power themselves.

Flower animals

Sea anemones are, despite their appearance, animals with a flower-like form; they are beautiful to look at but are nevertheless killers. Their tentacles are armed with venomous stinging cells, and small fish and shrimps touching them become trapped and then engulfed. Since the anemone's next meal is likely to come almost from any direction, the anemone's survival is best insured by a flower-like arrangement of tentacles around their mouth – a design that has been evolved secondarily by many bristle worms and feather stars.

A sea anemone is a wonderfully simple kind of animal, built like an open-ended sack. A basal disc attaches it to the sea bed or rock; and the mouth opens into a blind body cavity which is partially dissected by a series of mesenteries. A system of *circular* and *longitudinal* muscles are incorporated into the mesenteries and the body wall. There is nothing which passes for a brain, only a network of nerves.

The normal feeding response of swallowing the prey and rejecting the indigestible fragments involves relatively simple movements, but a few anemones are capable of quite complicated behaviour. One is a

fairly large species which is usually found perched on shells inhabited by the hermit crab; it is called *Calliactis parasitica* and occurs in British waters. The association is doubtless a mutually beneficial one. The hermit crab may well be safer from formidable enemies like octopuses by carrying around a stinging sea anemone, while the latter lives well by feeding on the crab's left-overs. However, the anemone's behaviour is geared to its rather precarious existence. It has to find a shell which is likely to be taken over by a hermit crab, then attach itself with all speed. Furthermore, it may fall off, or be left stranded when the hermit crab outgrows his present home and moves to larger accommodation.

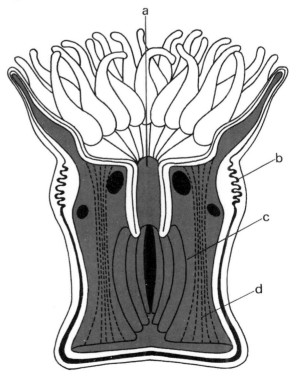

The anatomy of a Sea anemone. Mouth (a) – together with the sleeve-like valve and siphonoglyph. Main circular muscles (b). The gut partly divided by mesenteries (c). Longitudinal muscles (d).

A detached *Calliactis parasitica* reacts strongly to a shell by clinging to it with the tentacles and oral disc. Should the shell be occupied, the crab will often be held fast. The *foot* or the pedal disc then becomes detached, and the column flexes to bring the foot into contact with the shell. Only when the new attachment is secure will the 'head' of the sea anemone be released. Although not fast by our standards, this behaviour can be seen with the unaided eye. Scientific investigation has revealed that the 'parasitic' sea anemone responds not to the presence of a hermit crab, but to chemicals released by a part of the shell surface or *periostracum* laid down by the departed mollusc – a remarkable example of behaviour being triggered off by an intermediate signal messenger.

There is another sea anemone – *Stomphia coccinea* – which shows behaviour of equal fascination. *Stomphia* is usually found attached to the shells of the mussel, *Modiolus modiolus*. Unlike the home base for *Calliactis*, these mussels are themselves anchored to the ground by threads and so *Stomphia* has time to settle. Apart from finding the mussels in the first place, *Stomphia* itself may literally take off for it has a remarkable escape reaction from sea slugs and starfish. One kind of sea slug particularly is likely to feed on sea anemone tissue, and so *Stomphia* responds to the presence of the enemy by detaching its foot in an instant and wildly flexing the body which propels the animal through the water. Having escaped with its life, it has to establish a base on another mussel. When mussel shells are near by, the tentacles begin to explore, but when they touch one they do not cling onto the surface for dear life like *Calliactis*, for it has plenty of time. The foot is transferred by creeping movements rather than by performing an under-water cartwheel.

The hydrostatic skeleton

In order to understand the mechanics of their movement it is necessary to state something which is probably obvious to everyone, and yet is fundamental to the subject of animal design. Animals whether anemones or elephants are powered by muscles, and these only perform useful work when they contract. Indeed, muscles can *only* contract. They cannot expand on their own and so muscles are arranged either in pairs or groups that antagonise each other. When one set contracts, the opposing set is stretched, and vice versa. For example, in the vertebrate body the limbs are levers powered by flexor and extensor muscles; when the *biceps* contracts, as when we lift a pint of beer or tot of whisky to our lips, the *triceps* is stretched in readiness for powering the movement for replenishing the empty glass, which in turn will expand the biceps.

In sea anemones the circular and longitudinal muscles antagonise each other by squeezing on the volume of water enclosed in the body. Thus the body cavity acts as a *hydrostatic skeleton*. Water is incompressible and transmits pressure equally in all directions, so when the circular body muscles contract to make the body thinner, the longitudinal retractors become stretched as the body elongates. When the anemone is active, water is prevented from escaping by closing the mouth, and the sleeve-like throat or pharynx also acts as a valve. Nevertheless, even if the animal becomes

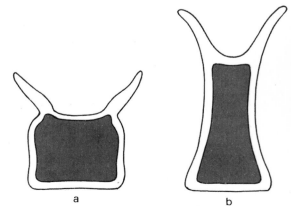

The effect of the longitudinal (a) and the circular (b) muscles squeezing on the body cavity.

Jewel anemones.

A sea slug, Archidoris pseudoargus. *This mollusc is not very closely related to the land slugs; the slug-form has evolved independently several times in the Mollusca. Note the fan of gills around the anus, and the sensory tentacles at the head end.*

A squid, one of the most advanced molluscan forms. Cephalopods are fast, highly active animals with comparatively large brains.

The design of the molluscan shell is nicely fitted to the species' requirements. The limpet's shell is adapted to withstand the battering of waves; cone-like shells have been evolved independently in many families that inhabit exposed shores. The greater the exposure, the taller the cones. Underneath, the broad foot acts as a sucker. Note the secondary gills and the fine sensory tentacles around the margin of the mantle.

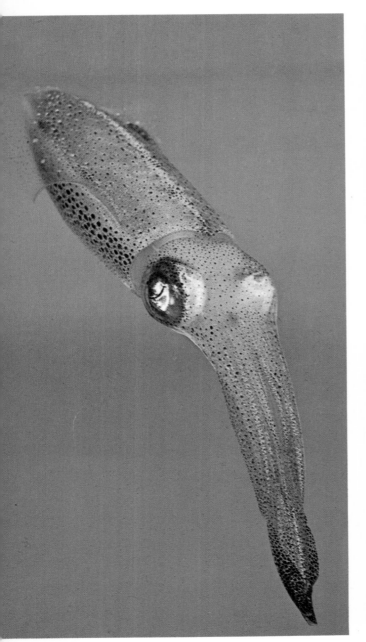

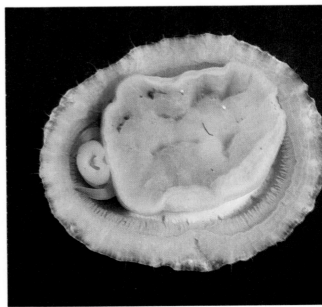

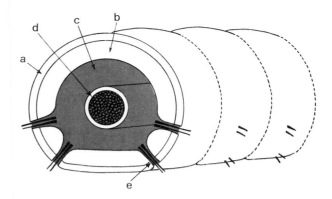

Diagrammatic transverse section of an earth worm showing the relationship between the circular (a) and the longitudinal (b) muscles, the coelomic cavity (c) and intestine (d). Each segment has four pairs of bristles (e).

completely deflated, the gut can become filled with water once again by means of a groove (the *siphonoglyph*) running from the mouth to the bottom of the throat which is strongly ciliated. Fine hairs waft a current of sea water into the body and can maintain a slight positive pressure equivalent to a head of two to six millimetres of water. It is not much, but it is sufficient to overcome the viscosity or 'stiffness' of the muscles, therefore stretching them and the gelatinous parts of the body wall (*mesoglea*). Once inflated, the siphonoglyph maintains a trickle charge of water into the gut to keep up the animal's turgor and to counteract the elastic forces of the body wall which would otherwise tend to shrink the animal.

The system described above is the hydrostatic skeleton at its simplest, and holds for the tiniest and most delicate hydroids, and corals, and the largest sea anemones which may reach a yard across. However, the single chamber of anemones has one drawback, namely that contraction of any muscle or part of a muscle will result in a change of posture of the whole animal because pressure will be transmitted in all directions. More sophisticated developments of the hydrostatic skeleton are found in worms, which will allow local alterations of body posture by the strategic positioning of pressure bulk-heads.

The worm at work

Those who are of the opinion that earthworms are hardly worth consideration should think again. In the soil they form a mostly unseen work-force of astronomical proportions. There may be three million to the acre on old grassland, and a density not far short of this beneath a well-kept lawn judging by the concentration of worm casts after a damp autumn night. No lesser man than Charles Darwin turned his mind to earthworms and, by weighing their casts in certain plots, reckoned that ten tons of soil per acre are brought to the surface each year. By spreading a layer of chalk and ashes over experimental areas in 1842 he

found that thirty years later they were covered by about seven inches of soil. He concluded that the surface soil over the whole country has passed through the gut of earthworms many times and, through the action of these soil workers large stones, together with our heritage of ruins, are gradually being undermined and are sinking beneath the surface at a rate of several inches every century. The remains of plants are dragged beneath the surface, broken up, and digested, and thus re-cycled by the worms which aerate and drain the soil by virtue of their ceaseless burrowing, for that is what earthworms are superbly able to do.

Burrower extraordinary

Earthworms have long round bodies, sharply pointed at one end, and their clean shape is devoid of permanent protrusions that would hinder movement beneath the surface. The body is divided up into a large number of segments in which many of the organs are duplicated. If an earthworm steak is taken and examined underneath a microscope, three dominating features can be observed. Firstly, the gut is plainly visible no matter where the section is taken, for it passes right along the length of the worm. Secondly, the body wall itself is very thick and muscular. On the outside, just beneath the skin, there is a thin layer of circular muscles enveloping powerful blocks of longitudinal muscles. Thirdly, and most significant from our point of view, between the gut and the body wall musculature is quite a large fluid-filled space that is technically known as a *coelomic cavity*; this forms the worm's *hydrostatic skeleton*. Unlike the sea anemone's, the coelomic cavity is incorporated into the body of the worm, and, furthermore, the bulk-heads or septa which divide one worm segment from another also break up the hydrostatic skeleton into a series of self-contained units. The two sets of muscles antagonise each other by squeezing on the coelomic cavities, which are as important for movement as the anemone's

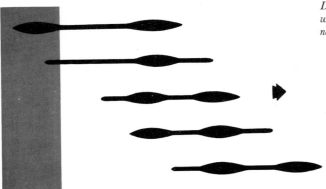

Diagram showing how an earthworm moves forward by passing waves of contraction of the longitudinal and circular muscles alternately down the body.

bag of sea water. A deflated worm, that is with the coelomic fluid experimentally removed from the spaces surrounding its gut, cannot crawl very effectively and can hardly burrow at all. A flaccid worm is a feeble worm, because the power of a hydrostatic animal is partly dependent upon the pressure it can generate in its fluid filled spaces by the muscles.

An earthworm moves by passing down the body alternate waves of contraction of the longitudinal and circular muscles respectively. The former make the body segments short and fat, while the circulars squeeze the segments out long and thin. The fat segments anchor themselves by the protrusion of fine bristles (there are four to each segment), while forward movement is restricted to those parts of the body where the circulars are active. When beneath the surface the fat sections are jammed against the sides of the burrow, the sharp front end can exert a considerable forward thrust.

Two features of the earthworm's design need explanation; firstly, the rôle of the septa, and, secondly, the disparity between the sizes of the circular and longitudinal muscles. By inspection it is quite apparent that the muscles making the earthworm increase its girth at the expense of its length (i.e. the longitudinals) are by far the most powerful; indeed, the longitudinals can exert a pressure ten times greater than that of the circular. The septa, by containing pressure differences in the hydrostatic skeleton in different parts of the body, allow the earthworm to use its longitudinals to full effect in burrowing without jeopardising the use of the circulars in other regions of the body. The absence of septas under these circumstances would otherwise cause the circulars to give way in the 'thin' parts of the body because the pressure in the confluent coelomic cavity would be transmitted throughout the whole of the worm.

Why then are the circular muscles not stronger? Probably because the burrow walls help to reinforce them when the longitudinals contract. One might also think that it would be in the worm's interest to

have strong circulars at the front for thrusting the body through the soil. In fact, earthworms probably move forward chiefly by seeking out existing cracks and crevices and then widening them by radial forces. Not surprisingly, it is the longitudinal muscles that would be responsible for making the body thicker and forcing the soil apart.

A further characteristic of burrowing worms is that they all tend to be round in cross section. The reason may be that in soft-bodied creatures, which have pressurised hydrostatic skeletons, the most efficient design is tubular, otherwise energy would be wasted in preserving a deformity. (Note that because of the way pressure is transmitted equally in all directions, a balloon that inflates like a discus would be difficult to make!). Leeches, for instance, are flattened for swimming, and in these parasites, special muscles are used to help maintain their shape. But, by and large, in engineering terms, a round worm is a mechanically efficient worm.

Hydraulic animals

Many examples can be found of animals which squeeze fluid in one part of their body to produce a corresponding force or change in shape in another part.

One of the most remarkable hydraulic systems built by nature is found in the legs of Echinoderms – sea urchins, starfishes, feather stars, and sea cucumbers. There are many jokes about centipedes and their proverbial hundred legs all kicking and tripping over each other, and yet a common starfish may have thousands, which all move in beautiful co-ordination.

All of the feet are hollow, and are plumbed into a single water vascular system which is charged from the sea via a perforated sieve plate or *madreporite*. Pressure of fluid in the ampulla antagonises the longitudinal muscles in the column of the foot; these are responsible for the walking movements. When the tube feet are retracted, the fluid inside is taken into the

ampullae; by shunting the fluid into the tube feet, they can be made to extend.

Spiders, too, operate by a combination of muscle and hydraulic power. They have no muscles to straighten out their legs, so they pump fluid into them instead. Muscles inside the body cavity are able to raise the pressure of the body fluid sufficiently to allow this force alone to antagonise the leg retractor muscles. In some ways the leg action can be compared to the party toy which uncoils when you blow into it, and springs back when the pressure is relieved; the spring represents the leg retractor muscles, which power the animal, and the air pressure the body fluid which straightens out the leg. Whether such insight will bring comfort to those to whom the sight of leggy house spiders in the bath arouses palpitations is doubtful!

Molluscs: the plastic design

No account of animal design would be complete without mentioning molluscs. Apart from their shells, they are largely soft bodied creatures and derive much of their turgor by pumping blood into spaces between their muscles, which otherwise antagonise each other in the manner described for anemones.

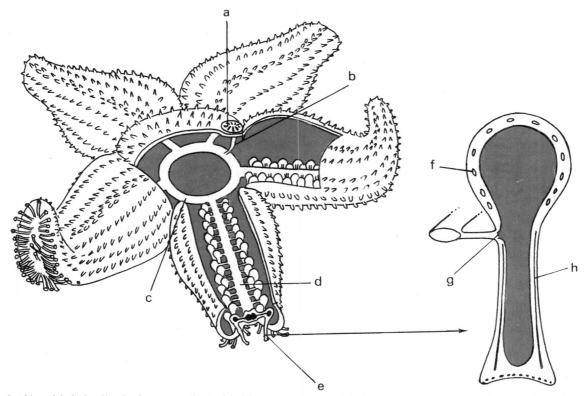

The plumbing of the hydraulic tube-foot system of a starfish. Water enters the sieve plate (a) and stone canal (b) into the ring and radial canals (c) and (d). Tube foot (e). Ampulla muscle (f). Valve (g). Foot retractor muscles (h).

15

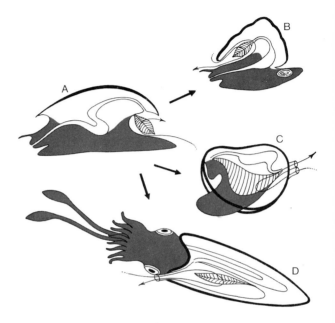

The Molluscs, along with the Arthropods, and Vertebrates are one of the most successful designs. There are at least 100,000 species and occur in the abyssal depths of the oceans and in deserts; some are anchored for life, others bore into the mud, rock or wood; many a good ship has been sent prematurely to Davy Jones Locker by the depredations of the ship worms which bored into their wooden hulls. Some are buried in the flesh of their hosts; squids torpedo through the sea, while many planktonic molluscs are carried almost passively by the ocean currents. The smallest snails could sit on the head of a pin while giant killer krakens are large enough to do battle with Sperm Whales – on whom they leave fearsome scars.

The basic molluscan design – insofar as there is one – has been moulded, modified, and manipulated by the process of natural selection to produce animals as diverse as mussels and octopuses, that look superficially as different as chalk and cheese! Indeed, one zoologist has described them as the all-plastic animals reflecting the extreme versatility of their basic form – and this is another characteristic of an evolutionary hit!

What is a mollusc?

A snail, 'shell-fish' and octopus must have many things in common to be placed side by side in the Mollusca. A mollusc is basically an unsegmented animal, that is, it shows no evidence of being constructed of a number of similar divisions like a worm. It has, apart from squids and cuttlefish, no hard internal skeleton and is bilaterally symmetrical (thus differing from starfish, sea urchins and their kin). The main features of the body are a muscular *foot* on which the animal moves which incorporates the head to some extent, and a *visceral mass* containing the digestive, excretory and reproductive organs; there is also a peculiarly molluscan 'invention' called the *mantle cavity*. The mantle enclosing the open ended cavity is a fold of the body wall. It contains the gills and also receives the waste and reproductive products. In aquatic species, water is broadly drawn in on one side of the mantle cavity where it is tested by a chemosense organ – the *osphradium* – then it sweeps over the gills, removing the faecal pellets and fluid excreta from the other side of the cavity. The edge of the mantle is responsible for secreting the shell if there is one, for not all molluscs have them.

The drawing above shows how the basic ingredients of the molluscan design are changed at least in emphasis to produce such widely different types.

In winkles, snails and sea slugs, the foot reaches dominant proportions; thus they are called the Gasteropods. Three-quarters of the molluscs belong to this class, and by invading the land they have shown themselves to be the most versatile of the lot. The foot is a highly muscular structure not unlike our tongue, and attached to the shell; when the columnellar muscle contracts, the foot is withdrawn into the safety of the shell, and can only be extended by forcing blood into the organ. It is a means of attaching the animal to the ground – tenaciously in the case of limpets – and

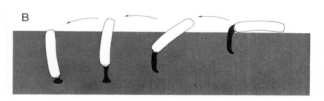

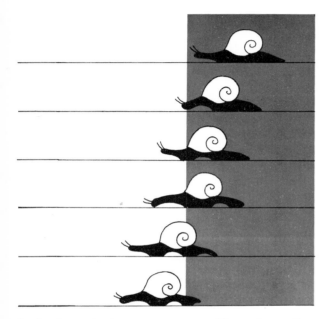

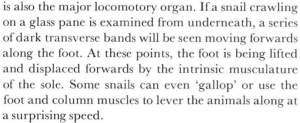

is also the major locomotory organ. If a snail crawling on a glass pane is examined from underneath, a series of dark transverse bands will be seen moving forwards along the foot. At these points, the foot is being lifted and displaced forwards by the intrinsic musculature of the sole. Some snails can even 'gallop' or use the foot and column muscles to lever the animals along at a surprising speed.

Most people associate Gasteropods with their shells, some of which are incredibly beautiful, both in form and colour; some rare cone shells are valued in thousands of pounds! It is basically a protective device for shielding the soft body against attack although admittedly time and time again the advanced members of different lines have dispensed with it (e.g. slugs). Several points are worth mentioning. Firstly, the structure allows for growth without change in shape or moulting as in the case of Arthropods. Secondly, coiling is usually a means of increasing the volume while retaining a compact form. It also makes for a very strong structure inside which the succulent

body is all but impregnable. The process whereby the crystals of lime are laid down is very well ordered and the coiling of any given species' shell is an example of mathematical precision.

In the bivalve shellfish or Lamellibranchs, the mantle cavity and gills reach their apogee of development. These animals are adapted to a fairly sedentary existence making a good living by sucking edible material from the surface of the mud or filtering off organic débris and plankton from the water, using the gills for straining and sorting. Some bivalves, like razor shells, are very active and burrow by plunging their foot into the sand, expanding it by gorging it with blood, and then contracting the foot retractor muscle. Using this method, they can burrow as fast as a man can dig.

High-speed squids and more stealthy octopuses have relatively enormous heads partly developed from the foot – thus Cephalopods. Their eyes and brain are second to none in the Invertebrate world, thus allowing an active and reactive mode of life. In Cephalopods, the muscles of the mantle cavity play an important rôle in jetting water through a funnel-like opening which can be pointed in any direction to propel the animal along. Some species can even launch themselves from the surface like rockets!

2

Living in armour

Insects, crabs and their kin

Introduction

There is something rather sinister about insects. Most are active industrious little animals which, on the face of it, show a modicum of intelligent behaviour. Some even form societies that have been compared with our own, and which appear to be run with military precision. More than one novelist has suggested that ultimately insects will inherit the world, and in the realms of science fiction the human race has periodically been threatened by monster ants and the like. If this seems too fantastic, the survival of millions of people is yearly threatened by voracious insects that eat our crops or which pass on dreadful diseases. From our own view-point, insect-like creatures are alien forms, which are built along totally different lines to those of our own bodies. They are encased in 'armour', and have fixed expressionless faces and more legs than we could cope with. The heart is an elongated, pulsing tube that lies where you would expect to find the spine, and their nerve cord runs along the bottom of the chest and belly! The way they sense the world is quite strange to us; an insect is likely to taste with its feet, smell with an antennae, and hear with its abdomen. We know that many can perceive ultra-violet colours beyond our ken. Some may sing but have no 'voice'.

The six-legged insects form part of a major group of animals without backbones, called the Arthropoda – meaning 'those with jointed legs'. It includes the chiefly aquatic Crustacea, and the predominantly terrestrial centi- and millipedes, the Arachnida (spiders, scorpions, mites and ticks), and Insecta, which constitute the largest class of all. By any standards, the Arthropods have been fantastically prosperous, indeed it is arguable that they out-rival even the vertebrates to which we ourselves belong if you take into account the number of species (a million or so), the amount of living material organised into arthropodan forms, and the effect they have on the rest of us. Also, the arthropodan design was capable of being adapted to survive and thrive in the most difficult environments on this planet, namely in habitats where water is all but absent (in deserts, and polar regions where water is locked away as ice); only the vertebrates and air-breathing snails have managed to conquer the land in a big way. Furthermore insects have mastered flight, the most demanding form of locomotion; only birds, bats, and pterodactyls have a similar claim to fame.

In the interests of speed and stability they have evolved legs for levering the body rapidly over the ground. Insects which are amongst the most advanced of arthropods have six so that they can always move them in two lots of three, two on one side and one on the opposite; they always have three points of support on the ground and are mechanically stable – like a three-legged stool. Their legs tend to splay out sideways, and in doing so resist side forces caused by wind or water currents which would tend to tip these light animals over.

What is of interest are the features of the arthropodan design which have allowed these animals to be successful and how these characteristics have at the same time imposed certain limitations. Here we shall concentrate on their external means of support and method of 'breathing'.

Living in armour

One of the most noticeable things about the adult insect body is that it has a soft centre and the outside is reinforced by a substance called chitin which is secreted in layers rather like plywood. This contrasts with that of a vertebrate which is soft and pliable outside but strengthened by an internal chassis of bones. Insects and other arthropods possess what we call an *exoskeleton*; that is, the supporting chassis is placed on the surface and derives much of its strength from its basically tubular lay-out and by its 'plywood'-like structure.

The exoskeleton, like a coat of armour, protects

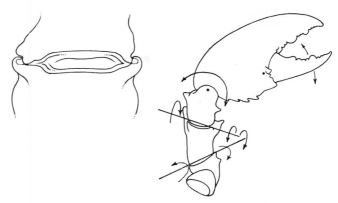

the soft tissues from damage and, in the land-living insects, spiders and millipedes, prevents the bodies from drying out – a very important job because small animals with an immense surface area in relation to their volume would tend to lose water extremely quickly. Body rigidity is also necessary for the legs to operate efficiently as levers, otherwise the internal locomotor muscles would tend to bend and flex the body rather than move the limbs. An arthropod's body cannot, however, be stiff all over; parts must be flexible enough to allow the joints to move – rather like a suit of armour. In these regions, the exoskeleton segments fit together by means of peg-and-socket joints; sometimes movement between consecutive parts of the legs are possible only in one plane, so that it becomes necessary to have two joints with opposing planes of movement placed next to each other to obtain complete freedom of movement. This arrangement is very easy to see in the claw-legs of crabs and lobsters.

In engineering terms, an exoskeleton can be regarded as a series of *hollow* cylinders; these are admirable structures for resisting bending and twisting forces of the kind that are likely to break bodies. It is therefore inherently very strong, for this reason: if you take a solid cylinder and try to bend it, one surface would be subjected to a *compressive* force, and the side that would be tending to split or stretch would be under *tension*. These opposing stresses and strains would cancel each other out in the centre of the cylinder, so strengthening material placed here is useless and can be dispensed with, without affecting the stiffness. A hollow cylinder is then the strongest structure for a given amount of building material. An insect's body is therefore tremendously strong for its weight.

There are, however, severe limitations to the use of armour for support and strength, which effectively places an upper limit to the practical size of insects. As cylinders become wider and larger, the walls would tend to become correspondingly thinner. Theoretically such a structure would retain its strength, but in practice it would tend to buckle and collapse when compressed rather like a tin can. This undesirable effect could be prevented by thickening the walls, but it would really require far too much building material to make this kind of construction a practical proposition. An insect the size of man would need an exoskeleton that would make the heaviest suit of medieval armour seem like tissue-paper if it were not to buckle all over! The arthropod method of support on the outside is by far the best for small creatures, but it does impose a limitation on the maximum size they can reach on mechanical grounds.

There are further disadvantages which are just as important. An arthropod must be able to grow and this is accomplished either by having a soft-bodied growing phase like a caterpillar, or by the expedient of periodically casting off the hard outer case, and putting on weight rapidly while the skin is laying down a new outer coat. For a period of a day or two, the new chitinous material is soft and stretchable, but then it becomes tanned or calcified after which further growth is prevented until the next moult. At this time the animal is greatly weakened, and so it must be light enough to prevent the new, flexible exoskeleton from becoming deformed during the hardening period. If insects grew over and beyond a certain size they would tend to collapse while moulting and harden into a shapeless mass!

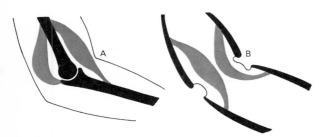

The difference between an internal or endoskeleton (A) and an exoskeleton (B) showing the layout of the muscles.

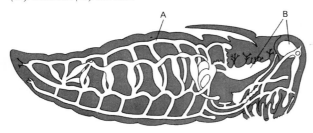

Diagrammatic layout of the tracheal system of a grasshopper. (A) Trachea. (B) Air sacs.

It is also possible to prove that an exoskeleton is highly vulnerable to impacts, and can be severely weakened by scratches. Again, these considerations are not important for small animals, but a very large insect would tend to break up very easily if it blundered into something. A large vertebrate chassis has the advantage of being cushioned by skin and muscles that can absorb the shock of considerable impacts.

Getting enough air

Apart from structural problems, increasing size does pose all manner of other difficulties, one of which is how to bring sufficient quantities of oxygen to the tissues where it is needed. Minute animals such as single-celled Protozoans or tiny worms receive all the oxygen they need diffusing through the tissues; every part of the body, even the deepest regions, are only a fraction of a millimetre from the air or oxygen-saturated water. Such animals have a tremendous area exposed to the outside relative to their bulk. Large animals run into problems; firstly, their surface area is reduced relative to their bulk thus exposing relatively less skin through which oxygen can enter the body, and, secondly, the surface tissues will tend to use up all of the available oxygen at the expense of the deeper layers. It has been calculated that the maximum radius a worm can reach if it relys solely upon diffusion is around about 1 millimetre depending upon its oxygen consumption. One development which can alleviate the situation is to have a blood system which helps to take oxygen from the surface of the animal and distribute it quickly to the deep-lying muscles; this raises the potential size of a skin breather to a radius of 0.3 centimetre or so, which is the size of the fattest South American earthworms.

Insects have evolved a system of delivering air directly to the tissues by a system of pipes or *trachea* which open to the outside via a series of port-hole-like *spiracles*. They take advantage of the fact that gaseous diffusion of oxygen along these passages is about 800,000 times as fast as the movement of dissolved oxygen through tissue. Only a small proportion of the insect consists of air sacs and tubes; even so, the theoretical size for a caterpillar using 0.3 cubic centimetre of oxygen per gramme of tissue per hour jumps to a radius of 0.9 centimetre – a creditable improvement on worms.

Some adult insects – like moths and dragon-flies – are very powerful and need a great deal of oxygen, particularly in their flight muscles. These are supplied with very large trachea and the muscles and air pipes are arranged in such a way that the muscle fibres are all within very easy reach of oxygen. A good supply is also facilitated by the pumping action of the flight muscles which cause a steady flow of fresh air to pass through the trachea.

It is doubtful whether this system of breathing is any less efficient than our own for small insects. If some form of thorough ventilation could be evolved, there seems no reason for suspecting that the tracheal system of oxygen transport would not suffice for insects larger than the biggest beetles of today. These possibly represent the maximum size the overall insect design can achieve in competition with smaller vertebrates; the bulkiest insect in the world is the Hercules beetle of South America which has a body length of 6 inches, and the slightly smaller Goliath beetle of Africa weighs about 3 ounces. Of course the crustaceans achieve much greater sizes because they not only are supported by water but also retain gills to help 'breathe' – structures which the land-living insects have dispensed with in favour of an air-plumbing system. Japanese spider crabs have bodies 12 inches across and their legs may span 12½ feet, although the North Atlantic lobster has reached over 40 pounds!

All in all then, it seems that the wildest nightmares of science fiction will never come to pass. Super-ants are mechanically just not on. However as an alternative design to the vertebrate one, the arthropodan or insect form is as good as any, providing sheer size is not taken as a criterion of prosperity.

3

On being a parasite
by Martin Wells

What is a parasite?

A predator is an animal that eats other animals, generally those smaller and weaker than itself. We respect predators, paint them on our coats of arms and use their names as adjectives of quality. To be termed lion-hearted, cat-footed and as strong as a bear offends no man. On the other hand, a parasite is an animal that eats other animals that are always larger and more powerful than itself. We despise them, and to call a man a louse, which is, after all, a courageous little animal in its way, is to ask for trouble. Our attitude is the more curious, when one comes to think of it, because so many of our own most successful activities come close to parasitism. The life of a herds-man, with milk or blood meals as a staple diet – which after all, is the living of whole races of mankind – is not in the end so very different from the life of the ticks and mosquitoes that share his source of food. All three survive because they manage to live off income, feeding at the expense of other animals with-out actually killing them. An outright predator, in contrast, is obliged to live on capital in order to remain alive. But even the 'income-capital' division is not a very satisfactory way of separating predators and parasites. It is merely the scale of the investment that differs; in the long run, predators must live on in-come too, or they will eliminate the species on which they feed. Their income is the excess breeding capacity of the stock on which they feed.

It all boils down to a matter of size. Parasites differ from predators in that, because of their relatively small demands, they can feed inside or on the surface of other animals without actually killing them. They always, however, injure their hosts to some extent and they give nothing in return for the food that they extract, a definition, incidentally, that lets out the herdsman because he does after all make some effort to protect the stock that he feeds upon.

There are believed to be many more species of parasite than non-parasite. Every host species has a number of sorts of organism, ranging from bacteria to tapeworms, that are more or less dependent upon him, animals that rely upon him to convert, directly or indirectly, the raw materials from plants into solid animal nourishment. With all that first class protein wandering about, it is hardly surprising that almost every phylum of animals includes creatures that have evolved means of exploiting an abundant supply of ready-made materials rather than go to the trouble of synthesising their own products for growth and energy.

Surface parasites (Ectoparasites)

The difficulty, of course, is that animals, even common animals, are pretty thin on the ground. The parasite must find its host and this may not be easy because the host is, by definition, larger and therefore generally swifter moving than itself. If the parasite detaches from the host after it has fed, the problem recurs each time it feels hungry.

There are comparatively few parasites that leave their host once they have found one. They are nearly all insect animals that succeed in this sort of existence because they are mobile and well instrumented. A mosquito or a tsetse fly can detect its host at a distance because it smells or sees him; a tick can climb into likely places to grab at the fur of passing animals; a bed bug can scurry away to hide in the mattress once it has fed. Outside the arthropods only the leeches have, as a group, adopted an intermittent, blood-sucking livelihood and even here the truly parasitic species are probably in the minority. Most leeches feed on snails and the like, and they very commonly kill their hosts at the first meal, a state of affairs that clearly smacks of predation.

Most parasites, once they have found a host, stay there. It means that they can afford to eat little and often, which is all to the good, since the habit of making a massive meal when the chance occurs leaves a tick or leech bloated and more than usually

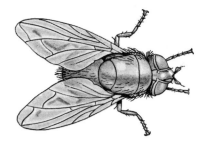

vulnerable just when it should be preparing to moult or lay eggs. But while staying put is fine for the individual louse that succeeds in colonising a new man or bird, it does little to solve the long-term problem of survival of the species. Ectoparasites staying permanently on their hosts manage in various ways. They may transfer from one host to another when these come into contact; crab lice, which suck blood, from a prostitute to her clients, bird lice, which feast on bits of feathers, from a hen to her chickens. Or they may have stages in their life cycles when they leave the host, which will tend to distribute the species because the available hosts may have changed by the time the infective stage is reached again. Fleas do this. The eggs drop out from among the hair or feathers of their host and hatch into little grub-like larvae. The larvae, unlike their bloodsucking parents, live on organic rubbish, particles of skin, left-over food, even the excrement of their parents, that they come across on the floor of their host's lair or nest. In due course they pupate and on emerging from the pupa after a period of reorganisation spread to whatever host or hosts is now using the amenities available. Since both larvae and adults can survive for long periods without food (upwards of 500 days for the human flea, for example) the odds are favourably disposed towards larvae hatching from eggs dropped in a nest. Eggs dropped elsewhere are unlikely to be as lucky.

But they are not, of course, quite as likely to be dropped outside. For animals, on the whole, tend to scratch and clean themselves at home, so an undue proportion of the eggs will be shed there. Indeed, matters seem to be even more biased in favour of the flea, because in at least some cases, and perhaps generally, the flea's reproductive cycle is geared to that of its host. Among rabbits, it has been found that the oestrogen level in the blood affects not only the ovaries of the rabbit but those of its fleas as well. Both become pregnant together, and the eggs are laid at about the same time as the mother gives birth. In due course each young rabbit leaves the nest with its own complement of young parasites.

Animals with fleas scratch and although this doesn't appear to worry the fleas much, the possibility of dislodgement is evidently serious enough to have favoured the evolution of a number of forms that go to considerable trouble to select places where the host cannot readily scratch or preen itself. One such is *Tunga penetrans*, the jigger, a flea with a life history like that of the common dog flea or human flea. The animal has a grub-like larva, and an adult that feeds on men, pigs, dogs, cats or rats – the jigger is not a fussy animal. Where it represents, as it were, an improvement on the normal range of fleas is that the females dig themselves into the skin. They select places like the edges of the toe-nails or the undersides of the joints of the toes to do this and once dug in are exceedingly difficult to dislodge. The female feeds, swells to the size of a pea, and becomes eventually little more than an egg bag with six legs, now useless and buried beneath the skin of the host. Only the tip of the abdomen projects, dripping eggs on to the ground as the host hobbles painfully about. Attempts to dig or rub out the parasite may cause the death of the flea but in any case leave an unpleasant wound that is well sited to collect any dirt that is around. People can die from the blood poisoning that follows an attack by jiggers.

Internal parasites (Endoparasites)

From everybody's point of view the jigger is a far from perfect parasite. It irritates and it gets killed. It is liable, inadvertently, to slay its host. Most parasites can do better than this and the jigger, as a species, is only part of the way towards evolving the capacity to do the job properly. The hosts of better adapted species are not taunted to potentially suicidal acts by the irritation produced by their parasites.

Among these better adapted forms is a wide range of organisms that have come to live actually within the tissues of their hosts. It is not hard to see how this state

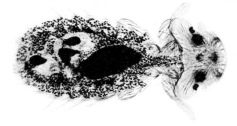

of affairs may have arisen. Invertebrates, like the jigger, show one way. An alternative entry is through the gut. Small animals and their eggs, inevitably, get swallowed by larger animals. If they can lodge in the mouth, or survive the digestive juices of their host, they find themselves in a well-protected and well-provisioned environment. In some cases this is all to the good of the host, since the intruder may be able to digest things that he cannot. Most herbivorous animals cannot in fact digest the cellulose cell walls of the plants on which they feed. They do it indirectly, by means of bacteria in the gut. In other animals the same good offices are performed by billions of single-celled animals (Protozoans). Termites cannot themselves digest wood, but have guts populated by the Protozoans – *Trichonympha* and its close relatives – that do the job for them. A termite fed on antibiotics so that it looses its gut fauna dies of starvation in the midst of plenty!

In contrast to these instances of 'symbiosis', where both parties appear to benefit from the association, are a great many cases where the relationship is clearly one sided. These range from tapeworms, which are relatively innocuous in that they merely extract a toll from food passing down the host's gut, to much more pernicious creatures like hook-worms that have taken to feeding on the host's gut wall instead. And from thence inwards, to settle in the blood and tissues.

Some of these internal parasites are among the most beautifully adapted of all creatures. Very highly specialised to their particular ways of life, they provide elegant examples of the extremes to which living material can proceed. They are also very important, economically because they attack the animals on which we feed and personally because a good many species attack us.

Animals living in the gut of man

Ascaris lumbricoides is a round worm and it lives in the gut of men and pigs. Nobody is quite certain whether those found in the two places are the same or different species of *Ascaris* and nobody, so far, has been inclined to try the obvious experiment. It is large as round worms go, with females a bit bigger than the males, ranging up to twenty or thirty centimetres in length. No great harm results from the presence of a few of these worms in the gut, though it can be rather alarming to pass them. Very occasionally, the animal penetrates into other tissues and it is then potentially dangerous, simply because it is large and disruptive. *Ascaris* copulates and lays thousands of eggs that pass out with the faeces. The eggs must be eaten for infection to occur. This sounds fine, since with adequate sewage disposal, there should be no problem. And there would be no problem, but for two factors. One is that very few parts of the world have adequate sewage disposal, for human manure is too valuable as a fertiliser to permit it, quite apart from anything else. The other is that the very small eggs are almost indestructible, so that even with the most elaborate precautions they tend to get back into circulation when the purified effluent from sewage schemes is returned to the rivers or used to water the land. A surprising number of people, even in Europe and North America where sewage is expensively treated, are infected. *Ascaris* is one of the world's common animals.

Also a nematode, and more serious because it gnaws the inside of the human gut, is *Ancylostoma*. A smaller animal than *Ascaris*, about one centimetre in length, its life history is similar until the eggs are shed on to the ground. In this case, instead of waiting until they are eaten, the eggs hatch into minute larvae that feed on the excrement with which they are dropped. They grow here and in the soil, moulting three times; having a hard cuticle, like arthropods, nematodes are obliged to moult in order to grow. The third-stage larva is infective. It doesn't feed, but hangs about on the surface of damp soil waiting for a host. If it is lucky it is trodden on and gets a chance to burrow into the skin of a new host. The skin itches, the host scratches, but it is too late, the worm is already in the bloodstream. It doesn't settle there, however, but circulates until it jams in the thin capillaries of the lungs. From there it breaks out into the air sacs and is swept upwards by ciliary currents and the host's coughing. The larva is swallowed and at last arrives in the hind gut where it settles down to gnaw its way to maturity.

Again, the odd worm does nobody much harm. But en masse hookworms are dangerous because of the amount of blood that they consume. An individual hookworm can devour about 0.5 cubic centimetre of blood a day. Infections of 500 animals or more are not uncommon and while the loss of quarter of a litre of blood per day may not be serious for a well-nourished man in the best of health, it is quite another matter for children, pregnant women or for anyone on the borders of anaemia because they are undernourished or iron-deficient. It has been estimated that nearly 500 million people in the world are infected with one or another of the common species of hookworm.

Blood and tissue parasites

Apart from the Nematodes, there is one other group of animals which appears, as it were, to be predisposed to parasitism. The Platyhelminthes, like the round worms, include many free living organisms – the flatworms which lead more or less respectable lives. But these creatures have a host of unsavoury relations, the flukes and the tapeworms.

One example will serve to show the sort of creature that we have to contend with. *Schistosoma (Bilharzia)* has proved to be one of the outstanding success stories of recent times. It is, for reasons that will become apparent below, an organism that profits from progress. It is spread by man travelling (so that it is now, for example, prevalent in large parts of New Zealand, where it has succeeded in establishing itself during the last fifty years) and it is particularly benefited by

irrigation schemes. Attempts to stamp out malaria by drainage schemes or to improve transport facilities by means of canals tend to benefit *Schistosoma* as much as anybody else.

All this arises because *Schistosoma* is a parasite of both men and of water snails. It infects the two during different stages in its life history and is spread by the movements of either. As an adult, the animal, a trematode fluke, lives in the blood vessels of man (one species *S. japonicum* infects domestic animals, rats and mice as well) where according to species it settles either in the walls of the hind gut or close to the bladder. The sexes meet, settle together and produce large quantities of spiked eggs that rupture the walls of the capillaries surrounding gut or bladder and escape with the faeces or urine. Once again, the damage done depends on the number of parasites. Individually small (1–2 centimetres), the effects of infection are progressive and debilitating.

The eggs must reach water to hatch and the ciliated larva that emerges must find a freshwater snail belonging to a small range of genera (*Bulinus, Physopsis* and one or two others) if it is to survive. If it does find a snail, it bores into it and begins to feed. Inside, it divides, producing a mass of small larvae that in turn divide to produce a brood of yet another larval form, cercariae.

The parasite is multiplied enormously and a single ciliated *myracidium* may ultimately produce some tens of thousands of descendants, which is rotten for the snail and mankind and splendid for *Schistosoma*. The cercaria larvae bore their way out of the snail and swim off. This is the infective stage, so far as we are concerned. Bathers, farmers in the rice-fields or fishermen with their nets, the women doing their family wash and anyone drinking the unfiltered water is liable to infection, so that in regions where *Schistosoma* is present, most people eventually get it. The cercaria larva burrows into the skin of its host, loses its tail and travels with the bloodstream until it settles, mates and begins the cycle anew.

Diagrammatic life-cycle of the fluke Schistosoma *showing how two hosts are involved: Man (A) and a snail (B).*

Which came first, the snail infection or ours? Is this organism to be regarded as primarily a parasite of snails, which has later in the course of evolution discovered a further and more nourishing host? Nobody knows, but it is suggestive that while the invertebrate hosts of trematodes are very nearly always snails (a few bivalves are known, and one polychaete worm) the final host may belong to any group of the vertebrates whatever. This would tend to support the view that the flukes are, in origin, snail parasites, for it is otherwise difficult to imagine why they should be restricted to a single class within the invertebrates from amongst all those potentially available.

It is, in any event, far from clear why such complex life histories should have arisen at all. There are others, even more elaborate, involving a sequence of no less than three successive hosts (see, for example, *Diphyllobothrium* below). One might well think that any benefits arising from multiplication in the larval stage would be more than offset by the much reduced chance of discovering a succession of suitable hosts in the right order. Apparently this isn't so; these animals manage very nicely, and one must suppose that the chances of successive infection are a good deal higher than would appear at first sight. We know, for example, that cerceriae tend to escape from their snail host at particular times of day and this may be a response to the habits of their targets. At the other end of the life cycle we don't know how the miracidia find their snails, but we can observe them clustering around suitable hosts and can guess that bumping into a snail is anything but a random process. Quite possibly the benefit of complex multi-hosted life histories lies in the establishment of two or more largely independent populations of parasite from which the species can be re-established should either or any of the host species become for a while rare and unavailable.

Effects of parasitism on the parasite

The behaviour of adult endoparasites is always simple. They sit and suck or absorb their food through their skins from the nourishment around them. Conditions remain constant because the host regulates its own internal physiology, and if conditions do change there is nothing the parasite can usefully do about it. Its job is to make the most of its opportunities while they last and it is the capacity to produce thousands of eggs that matters in the end. In consequence of all this one finds that endoparasites have few sense organs, very little nervous system and few if any organs for locomotion. Digestion and excretion are often a matter of absorption and diffusion of solutes, with the host doing all the rest of the work. The reproductive apparatus fills the body.

These are the structural adaptations. But it should not be forgotten that there are, as well, many less obvious adaptions that fit the parasite to its peculiar way of life. It must be resistant to the host's digestive enzymes, if it lives in the gut and to the host's antibodies if it lives in the blood or tissues. In the normal way, invasion by a foreign body produces a rapid onslaught from the defensive mechanisms of the body. Amoeboid *leucocytes* devour the intruder, antibodies are generated that stunt or prevent growth. In the normal way it is impossible even to transplant tissues from one mammal to another of the same species – the immune response leads to rejection of the graft. Parasites somehow get by. They somehow avoid stirring these forces into action and remain free to circulate in the host's tissues as if they were bio-chemically identical with the host. This doesn't always work; sometimes the parasite is attacked, and in some instances the host builds up a progressive immunity that eventually prevents further settlement by parasites of the same species. Snails, for example, tend to do this in response to flukes, so that some of the older members of the population prove to be the only specimens uninfected when they are examined.

Diagrammatic life-history of the tapeworm Diphyllobothrium latum, *showing how three hosts are needed to complete the cycle. Eggs (A) are passed in our faeces, and produce mobile* miracidia *(B). Asexual stages are found in Copepod crustaceans (C) and fish (D). We get infected by eating fish, and the adults (E) live in our intestine.*

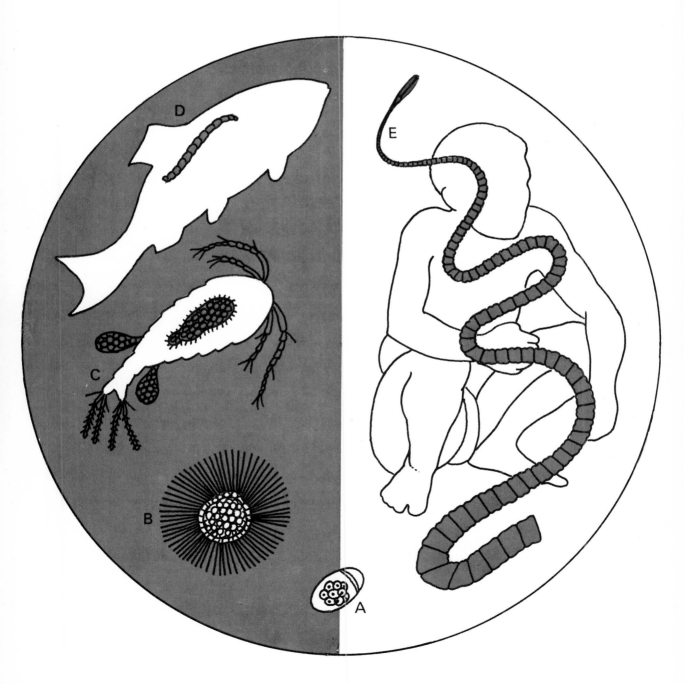

The very precision with which the parasite must match its host is one of the factors that limits its distribution. If it invades the wrong host it is liable to be slaughtered, which is just as well in view of the enormous numbers of parasites about; almost every freshwater snail will yield large quantities of cercariae if it is kept warm and well lit in an aquarium. It is fortunate indeed that our own immunity system is so specific and so difficult to match.

Tapeworms

Having hooked itself to the gut wall, the tapeworm, which is gutless and practically brainless, settles down to reproduction, producing prodigious quantities of eggs in a production line of packets, the *proglottides*. The eggs pass out with the faeces. Along the shores of the Baltic sea and in eastern Europe generally, man is at times infected with a giant among the tapeworms, *Diphyllobothrium latum*. The creature, which may eventually grow to a matter of ten metres in length, is obtained by eating raw or insufficiently cooked fish. But the fish does not itself collect the tapeworm eggs when they are dropped from man, as in the sort of life cycles already considered. It acquires its parasites by eating copepod crustaceans (*Cyclops* and *Diaptomus*) that eat the first-stage larva when it hatches from eggs voided with man's faeces.

Once again one is appalled by the difficulties that the parasite seems to set in its own path by having such an elaborate life history. The chance of a crustacean picking up a larva cannot be very great, the odds against the crustacean being eaten by a fish that is subsequently caught by a man (or a mink, or a seal) are, one might imagine, very long indeed. But all these events must happen often enough, because the animal survives and, indeed, is still spreading. *Diphyllobothrium* colonised the great lakes of North America between the two world wars. The answer must lie in the enormous reproductive capacity of the adult worm, for the millions of eggs that it lays, like the millions spawned by sessile animals, individually, have only an infinitesimal chance of surviving to maturity.

It is fine to be strong and swift and brainy. With enough elaboration of sensory, nervous and manipulative equipment it is possible to move mountains. We tend to think that because we can control our environment, more or less, this is the ultimate goal of all animal species, the end to which the evolution of animal types must inevitably proceed as a result of interspecific competition. The sessile animals show that it isn't and the parasitic animals, which, it is well to remember, probably outnumber all the rest of us, show that survival by no means depends upon the capacities on which we so pride ourselves. We think brains are important, because that is our speciality. It is salutary to reflect that there are plenty of other ways of making a living.

4
Fish form—the new look

Introduction

When reeling in a roach wriggling on the end of a line, or relishing the pink flesh of a salmon, one might be tempted to subscribe to the view that man is the superior being. In some respects, the notion is perfectly valid – although perhaps a trifle egocentric. A more intellectually satisfying thought might be that in some ways we are greatly indebted to them because our forbears were fish-like creatures, and even today we share the same basic design, one which appeared fairly late in time after insects, molluscs and sea-urchins and their kin had made an entrance. Originally, the fish-like design made for efficient movement in water, but its potential was much greater than probably any other basic animal form because it became modified to produce the amphibians and the land-based reptiles, birds and the mammals to which we ourselves belong. We all have a common ancestry and are built to a common plan, and together we constitute one of the major groups of living things called the Vertebrates.

One can expect to see the design at its simplest in the type of animals from which the root of the vertebrate family tree sprung. Surprisingly, we all probably evolved from such unpromising animals as sea squirts and their kin. These could be mistaken for sea anemones at a glance and live a sedentary way of life by filtering plankton from a powerful stream of sea water that is sucked through a net-like sieve in what could loosely be called a mouth cavity. From the appearance of these immobile animals one could never imagine anything further from our own line of accession. However, like so many 'fixed' animals, they have mobile larvae and those of sea squirts have all the makings of a respectable fish. The sea squirt larva looks rather like a tiny tadpole; it has a body, and tail projecting beyond the anus, a ventrally placed heart, a dorsally placed hollow nerve cord, and tiny 'gill' slits to boot. Most significant of all, running midway down the body, is a primitive kind of 'backbone'

called a *notocord*, on either side of which are arranged a segmented series of muscle blocks or *myotomes*. The flexible, internal notocord not only supports the muscles but also acts as a stiffener to prevent the body shortening when the tail sculls the tadpole forwards. The layout of the organs fundamentally differs from those of other animal groups such as the insects, molluscs and worms.

It is possibly from such lowly origins as these that the animals with backbones evolved – which is what the term Vertebrata means. Biologists have postulated that this would have involved the larvae of such animals as sea squirts developing reproductive organs before becoming sedentary – thus becoming free to breed while swimming about in the sea. From this stage it would have been a relatively short step to animals that we would recognise as fish and the changes would have been roughly as follows. Those individuals that developed more muscular tails would inevitably have been able to swim more quickly than their rivals, and the evolution of a segmented, strong backbone probably went hand-in-hand with increasing size and swimming power. Filter feeding probably gave way to grubbing nourishment from the bottom deposits of rivers, lakes and seas, and this innovation was possibly connected with the need for a greater intake of energy caused by a more active way of life; the 'gills' accordingly gave up their major rôle as food strainers and became predominantly concerned with absorbing oxygen from the water and getting rid of waste carbon dioxide.

The first fish were rather primitive creatures, often heavily armoured against the depredations of voracious invertebrate predators, like giant water scorpions (long since extinct relatives of the modern spiders) that abounded in those far-off days. However, they still lacked two important features; jaws and paired fins. The two great achievements of fish as we know them has been the development of these structures. There is every reason to suppose that jaws allowing the big bite, and the pre-processing of

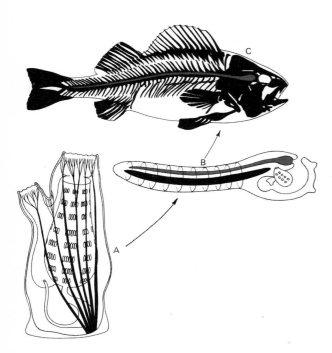

Diagrammatic representation of the evolution of fish.

(A) Adult sea squirt.

(B) 'Tadpole' of a sea squirt showing the relationship between the nerve cord (grey), notocord (black) and segmented muscles.

(C) The essential layout of a fish is in many ways similar to that of (B). The notocord has all but disappeared and been replaced by a bony segmented vertebral column which incorporates the dorsally-situated nerve cord (grey). The back muscles are omitted to avoid confusion.

food before it was swallowed, were evolved from a series of bones of the kind that support the gills. Paired fins strengthened internally by bones gradually appeared, their rôle – as will be described in greater detail below – was that of bringing greater control and manoeuvrability to swimming, but, of course, later they were to change into limbs and wings; the fact that we have two pairs of limbs, and not more, can be directly related to the development of two lots of paired control surfaces in fish!

· It is interesting to reflect on the fact that the success of the vertebrates was by no means assured right from the beginning. At a time when the most advanced members of the vertebrate kingdom were heavy and rather sluggish jawless fish, an impartial observer from outer space might well have placed the odds heavily in favour of the water scorpions or Cephalopod molluscs, or even insects, inheriting the earth.

Fish, however, developed jaws; these enabled them to take in more fuel, and, in turn, allowed for more powerful swimming. They also had a basic design that was capable of tremendous development. So fish won the day, thereby making way for a line of vertebrates that were to change the face of the planet.

Streamlining

A fish, or any other body making headway beneath the surface comes up against a number of problems, which can be alleviated by assuming the right shape. Firstly, water is a dense medium that has to be displaced at the front of the fish, and filled in at the rear. In flowing round the body, the water is likely to be thrown into an underwater wake. Secondly, water is viscous; in other words, it tends to stick to the skin and causes *friction*. Both make themselves felt as *drag* and sap away energy. The evolution of streamlining reduces drag and allows movement through water with a minimum of effort. If there was an ideal streamlined shape that would cause least disturbance of the water, then fish would presumably all be very similar in outline. However, drag forces will depend upon all kinds of variables, such as the length of the fish, its cross-sectional area (i.e. the front that it presents to the water) and the speed at which it normally swims. A sluggard like a carp or goby has a different shape from a blue fin tunny that can slice through the sea at over forty miles an hour. By and large the sort of form that gives least drag has its greatest width about one-third of the way along the body, and is approximately four and a half times as long as its maximum width or diameter. A trout's body approaches these proportions. However, fast species like tunny and bonito have evolved a body form in which the deepest part is way back towards the tail. The reason for this is that, at high velocities, the boundary layer will tend to remain smooth or laminar as far back as the thickest part; by having a body shaped like an elongated wedge, turbulent flow,

and thus drag, can be kept down at high speeds. Friction between the skin and water is less significant at low speeds, but is a force of increasing importance as the speed rises. Not surprisingly, the fastest fish have very smooth skins, with the scales well covered, so as not to induce turbulence; compare a herring's with that of a fast mackerel's. The effectiveness of streamlining can be demonstrated by the fact that a tunny weighing about as much as a man can move ten times as fast as an Olympic champion swimmer!

Of course, streamlining is not necessarily the only consideration affecting the evolution of the basic fish form. If one contrasts the beautiful flowing lines of a blue shark moulded by the need for speed and efficiency with that of a sea horse, clearly the survival of the latter species is more dependent upon staying put and being camouflaged, and its shape obviously plays a part in concealing the creature. In the underwater jungles – the coral reefs and tangled environment of the algae beds – manoeuvrability rather than the ability to thrust through the water are factors

which have contributed to the design of the inhabitants. Many fish that lurk in crevices are long and thin like moray eels, and a whole array of bottom dwellers have tended to become flattened one way or the other. Skate, ray, angler fish and to some extent gobies, settle on their stomachs and are dorsoventrally flattened, while plaice, dab, halibut and their relatives lay on one or other of their sides (depending upon the species). Many of the peculiar deep oceanic fish have greatly enlarged mouths and much reduced trunks; in the darkness of the deeps, their sole concern is not wasting energy on movement: thus few swimming muscles – and not missing a chance of a meal should one float by: thus their 'fish-trap' jaws.

Fish power

A fish can only move providing it can generate a thrust that opposes the combined drag and frictional forces. In practice, this means that the animal must accelerate water *backwards*. This is achieved by flexing all or part of the body into a series of backward moving waves which thrust against the water and moves the fish forwards.

The lateral undulations are made by a nicely controlled pattern of contraction of the muscle blocks

Dog fish swimming, showing how the body flexes into a series of waves which pass down the body, thrusting against the water and causing the fish to move forwards.

The segmented arrangement of muscle blocks or myotomes in a fish (A). All but two have been dissected out to show their double cone-like structure (B).

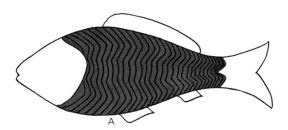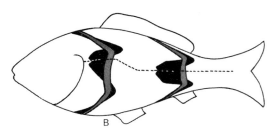

placed on either side of the backbone, and these constitute quite a high proportion of the weight of fish (this is why they make very good eating). Fast and powerful fish are much more muscular than the slower ones; goldfish have back muscles which account for only 20 per cent of their weight, whereas in trout the proportion approaches 66 per cent and no less than 75 per cent of the tunny's bulk is invested in its power plant.

One further thing deserves to be mentioned about the arrangement of the back muscles. Anyone who has eaten a sizeable salmon or cod steak will probably have realised that, apart from being segmented, the flank muscles have a double cone-like shape and consecutive muscles fit nicely into each other. The reason for such an arrangement is that the contraction of one muscle can exert a bending force over several segments of the vertebral column, and presumably makes for a better and smoother action – as well as extending the area of attachment between neighbouring muscle blocks.

The internal backbone is ideally constructed to withstand the alternating tensile and compressive forces caused by the contracting muscles, while retaining a necessary degree of flexibility.

Eels show a very basic kind of swimming because the rearwards travelling waves are of high amplitude and therefore very obvious. Fish in general, however, have dispensed with such flexible bodies and have modified their method of movement in the interests of higher efficiency. Generally, fish have restricted the area of undulation to the rear half or even the rear third of the body. This means that the backward-moving 'wave' starts somewhere in the centre of the body and by the time it reaches the tail, has a very high amplitude – in other words, the tail fin is wagged rather widely from side to side.

The thrust generated at the rear end will depend largely upon the amount of water projected backwards and the rate at which it is accelerated. The broad and slightly flexible tail (or *caudal*) fin operates by sweeping against a broad swathe of water and therefore increases the amount of thrust. Fish with their tail fins removed can still swim, although they lose a large proportion of their power, and thus speed.

There are, however, further refinements of design. Vigorous undulations of the body are likely to generate forces which are unbalanced and cause the fish to wobble. By restricting the lateral movements to the rear portion of the 'hull', many of these forces tend to be ironed out. Nevertheless, the fluctuating momentum of water, first on one side and then the other, may still produce a degree of body recoil, which would tend to impair the swimming efficiency by creating turbulence. This phenomenon is reduced by keeping the body very thin in regions where it is moving sideways against the water but not producing much useful thrust. The region just in front of the tail therefore tends to be narrow, and indeed extremely thin in fast swimmers which wag their tails at high frequency. Furthermore, any tendency of the front and stiffer end of the body to wobble around laterally or *yaw* in response to the vibrating tail region is resisted by the development of a deep forward section,

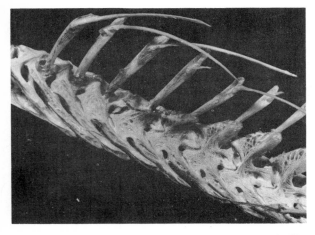

Longitudinal section of a fish's backbone showing the lattice-like structure which gives it strength.

32

The salmon (A) and rockfish (B) have typical heart-shaped caudal fins, whereas high-speed species like the tunny (C) and the marlin (D) have 'lunate' tails. Also note the relative positions of the paired fins. In (A) the pectoral and pelvic fins are in the fore and aft positions, but in the more advanced kinds, the pelvic fins move forward to lie beneath the pectorals.

aided by erectable *dorsal* and *anal* fins which effectively increase the resistance of the hull to unwanted sideways movement.

Before passing on to consider the subject of manoeuvrability, it may be of interest to see how different kinds of fish which are only distantly related to one another have all independently evolved similar tail forms in order to solve – as it were – the problem of achieving high speed without impairing a heavy drag penalty. Species of no great athletic ability possess tail fins which are broadly heart-shaped (A and B). They have an ample surface area resulting partly from the need for increased depth, because it is the depth that will determine the mass of water pushed backwards with each sweep. However, a tail of this construction will generate a great deal of drag at high speeds because of its surface area. High-speed fish need a generously deep tail fin to produce high thrust without too much surface area – in other words a high aspect ratio caudal fin.

By the process of convergent evolution a host of fast fish (and aquatic mammals!) have developed slim, sickle-shaped (or *lunate*) tail fins. For example, the tunny and marlin (C and D) are both voracious fish killers cutting through the surface waters of the oceans, vibrating their high, thin lunate tails at perhaps ten times a second. They have unusually high blood temperatures of 30°C or so, and in the mackerel family have even dispensed with pumping water through the gills; they simply ram water through as they swim along with their mouths open! Some sharks, even dolphins and propoises, and the extinct *Ichthyosaurus* have lunate tails.

Steering and stopping

Movement without control is no good to anyone. A series of control surfaces have therefore been evolved so that forces which would otherwise tend to throw the fish off course can be checked. A fish may tend to

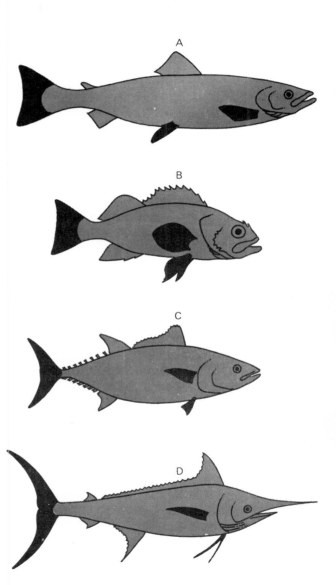

A

B

C

D

33

Most fish scull themselves along using their caudal fin (e.g. herring E). In many families, the other unpaired or paired fins have taken over the function of propulsion (wrasse F, pufferfish G, sunfish H, and seahorse I).

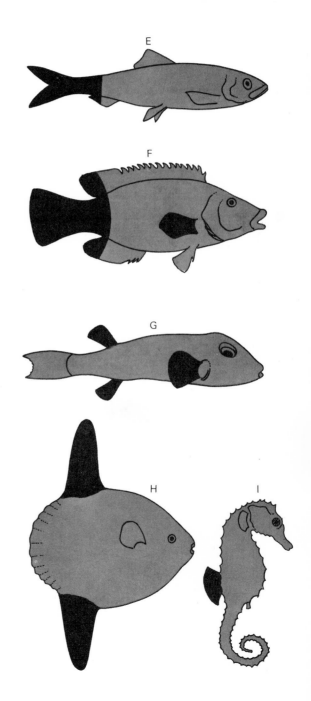

pitch on to its nose or tail, or *roll* about its longitudinal body axis, or to *yaw* from side to side. The control of pitching, rolling and yawing is achieved chiefly by the fins, which, in bony fish, can be erected or depressed.

Forward pitching is prevented by the use of the paired *pectoral* fins; by spreading them and inclining them at an angle to the water stream, they behave as hydrofoils, creating a lift which thereby rotates the body upwards around the fish centre of gravity (more usually called the centre of pressure). The hindermost *pelvic* fins, which in many fish are placed behind the centre of pressure, can make the tail rotate upwards to prevent rearward pitching.

The paired fins also double up as brakes. In the fore and aft position, as in salmon (A), there is a tendency for the pectoral fins to pitch the fish upwards when it wishes to stop, but in the more advanced bony fish, e.g. rockfish (B), the pelvics have moved forwards to a position just behind the gill covers, while the pectorals move vertically upwards to the mid-line. In the forward lower position the pelvics counterbalance the lift generated by the pectorals by the same kind of force that is likely to pitch a cyclist over the handlebars if he brakes too severely on the front wheel alone. The fish is therefore able to stop quickly in a stable manner by the use of its four paired fins.

Unpaired fins as well as the paired ones prevent rolling, and the former help to stop yawing (see above). Species which live by making a rapid dash for prey tend to have long thin bodies with dorsal and anal fins placed just in front of the tail fin. Rather like the vanes in an arrow, the posterior placing of these fins doubtless help the fish to swim on a true, accurate course. This arrow-like shape tends towards high stability at the expense of manoeuvrability. On the other hand, a great many highly manoeuvrable fish have short deep bodies with large fins which can be raised or lowered to swivel the body one way or another.

Not all fish have supple, flexible bodies, powered by a vibrating tail. In many families, the paired fins

themselves have taken over the function of propulsion, turning the caudal fin into a rudder. The gorgeously coloured wrasses flit through the marine gardens like so many birds, using their pectoral fins like wings. The analogy between fish and birds seems even more appropriate in the case of the cartilaginous rays and mantas whose motion on undulating, muscular pectoral fins is such a delight to watch. Some fish emulate birds in an even more direct way – the flyingfish. These have enormously developed pectoral and, to a lesser extent, pelvic fins; they dash through the water at perhaps twenty miles an hour, break the surface continuing to thrust with the much elongated lower spur of their tail until their wing-like pectorals generate sufficient life to launch the fish into the air. Once airborne, the pelvics are spread to provide added lift. By sallying forth perhaps several hundred yards into the air they are able to escape their fish and dolphin enemies – although they may become vulnerable to birds in doing so. Mudskippers come up for air and insects on their paired fins which are modified accordingly to act as limbs. Others scull themselves along with their median fins; the sea horse's fan-like dorsal fin provides most of its power while the curious oceanic sunfish has all but lost the tail, and rows along on enormously developed anal and dorsal fins. A walk round any aquarium will provide a large number of examples of varied forms of fish propulsion.

Fins for different jobs

Apart from taking over the responsibility for producing movement, there has been a considerable degree of adaptive radiation in fin form and function. In the gurnards, some of the pectoral fin rays have been generously endowed with 'taste' buds; when hunting, these fish 'walk' along the bottom on these greatly enlarged fin rays, which can detect food 'underfoot'. Numerous species which inhabit the intertidal regions of the shore have evolved sucker-like pelvic fins which enable them to sit tight. The lure above the angler fish's mouth has developed from a dorsal fin ray. Others, like weaver fish – have spiney dorsal fins associated with powerful poison glands. As a general rule, animals which are capable of defending themselves adequately usually display the fact by being conspicuous. Fins play their part.

The fantastic development of fins in the dragonfish exceeds the hydrodynamic requirements of the animal (like the tail of the peacock jeopardises aerial manoeuvrability), but together with the colours warns potential enemies of the venomous nature of the fish. The erectable nature of fins makes them particularly suitable for signalling. Of particular interest are the egg-dummy patterns situated in the anal fins of certain Cichlids. *Haplochromis* is a mouth brooder; the female lays the eggs and then takes them into her mouth until they hatch. The male fertilizes them by dropping his anal fin displaying the egg dummies and shedding his milt at the same time. The female reacts by attempting to retrieve the 'egg dummies' but at the same time taking the sperms into her mouth, where the eggs are lodged. In numerous species the fins have been modified by the demands of courtship.

Neutral buoyancy

The materials of which animals are built are, by and large, heavier than fresh or sea water. In other words, all things being taken into account, a fish composed of flesh and bone or cartilage will sink. This is an ideal state of affairs for species which habitually

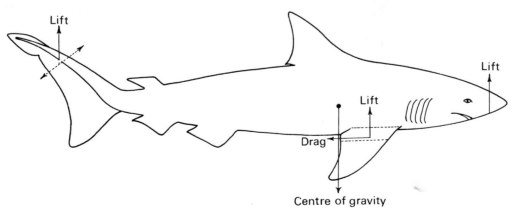

Forces acting on a shark when swimming. These fish are heavier than water and can only keep in mid-water by swimming. The asymmetrical tail, pectoral fins and nose create lift (see text).

Long-finned tunny or albacore – one of the fastest of fish. The front part of the dorsal fin and the pectoral fins fit into grooves when depressed to improve the streamlining.

Pollack (top left), Sea bass (below).

Swim bladder of pike opening into the fore-part of the alimentary canal.

lurk on the bottom, like a flounder or skate. The majority of fish conduct their business in mid water which means that they would have to expend energy in order to stop sinking.

This is precisely what most sharks have to do; a dead one sinks. In order to keep afloat a shark must always be on the move – a curious state of affairs, but which is nevertheless made easier by the 'design' of the animal. The shark's leathery fins are fixed in form, although the angle of attack of the paired ones can be altered. The front pair behave like hydrofoils and lift the nose upwards, thus keeping the fish afloat. However, a degree of vertical manoeuvrability in the heavier-than-water species is maintained by the evolution of an asymmetrical tail configuration. Lift as well as thrust is generated by the long upper lobe of the caudal fin, and this would tend to pitch the fish forwards into the depths. There it stays unless it can swim fast enough to generate sufficient lift by its pectoral fins to lift the nose upwards. Some of the larger sharks overcome this problem by storing oil in their vast livers; such species have almost symmetrical lunate tails.

Bony fish have become complete masters of their watery environment by largely becoming neutrally buoyant with the evolution of an internal buoyancy tank – or swim bladder. This is a thin walled sac developed from the fore gut, into which chiefly oxygen is secreted, the buoyancy of which compensates for the apparent weight of the fish in water. At any given depth, the fish can keep station without exerting any energy save that necessary for neutralising the thrust of water being pumped through the gills. Because sea water is denser than fresh water, marine fish can float more easily than their freshwater cousins, and so need rather less assistance. Accordingly, one finds that marine fish have bladders that are equivalent to 5 per cent of their body volume, compared with 7 per cent in river and lake inhabiting species. These figures correspond to the theoretical sizes of the swim bladders which are necessary for eliminating

the apparent weight of the fish in water.

Air, however, has one disadvantage as an aid to floating; it is compressible. A fish, theoretically, would become denser as it swam more deeply, and thus lose its ability to relax. Conversely, a deep-water fish, with a swim bladder filled with air under considerable pressure would likely rupture itself if it suddenly rose to the surface. These problems are largely overcome by the ability of fish to compensate for changes in depth; a richly vascular gas gland is able to secrete and absorb gas from the bladder so that, providing a fish does not make vertical excursions of too great a magnitude too rapidly, then the activity of the gland can adjust the volume of gas inside the bladder. If air is experimentally removed from the swim bladder of a goldfish, making it unable to float, the volume of the bladder is soon made good. Some deep-water kinds, like hatchet-fish, make daily vertical migrations of perhaps 500 metres, and these must be able to adjust the air pressure in their bladders between 5 and 50

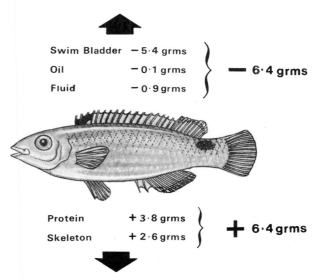

Swim Bladder	− 5·4 grms
Oil	− 0·1 grms
Fluid	− 0·9 grms

− 6·4 grms

Protein	+ 3·8 grms
Skeleton	+ 2·6 grms

+ 6·4 grms

Buoyancy budget of a goldsinny wrasse, Ctenolabrus rupestris *– a sea fish.*

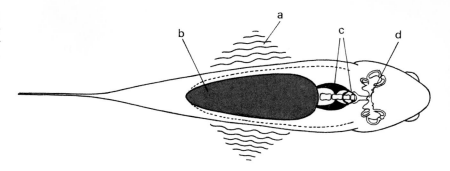

A longitudinal horizontal section through a carp-like fish showing how water-borne sounds (a) are conveyed to the inner ear (d) via the swim bladder (b) and Weberian Ossicles (c).

atmospheres! Admittedly, many bony fish have lost their buoyancy tanks, particularly extremely deep-water forms living below 1,500 feet and also some of the powerful oily mackerels.

The swim bladder, no less than the fins, has proved to be a very versatile organ. Apart from its hydrostatic function, it acts as an oxygen store; fresh water fish may be able to live for 10–15 minutes on their internal oxygen supply alone, and for deep-sea forms, a great deal longer. In many species, the swim bladder remains open-ended and can be used as a lung. This is particularly important for fish that live in muddy, swampy surroundings where the water often becomes depleted of oxygen. To be able to gulp down air under these circumstances is vital to survival. Lung fish, certain characins, mud minnows and the rather primitive bichirs, gar pikes, and bowfins use their swim bladders in this way for respiration.

Some zoologists think that the swim bladder originally evolved as an air-breathing organ, and that its function as a buoyancy tank is a secondary one. Whatever the truth of the matter is, the success of fish in coming out of the swamps to colonise the land depended largely on being able to breathe air, and the vascular swim bladder allowing fish to survive in oxygen-deficient waters was ready to take on the rôle of an amphibian lung.

Most remarkable of all must surely be the use of the gas bladder for detecting and making noises. There are, however, fundamental difficulties in sensing sounds beneath the surface because fish flesh is similar in density to that of water and is therefore transparent to the sounds – a problem that the highly vocal whales have also had to cope with. A gas bladder will tend to stop water-borne vibrations which will cause the walls to resonate. In Cyprinoid fish – a family that includes goldfish, dace and roach – vibrations of the swim bladder are transferred to the inner ear by way of a series of bones called the *Weberian ossicles* – named after their discoverer. These bear an incredible functional resemblance to the three tiny bones in our middle ear which amplify the vibrations of the ear drum and transmit them to the fluid of the inner ear. This arrangement makes Cyprinoid fish very sensitive to sounds; for instance they will respond to noises between 30 and 50 decibels lower than the sounds to which eels will react. This compares well with the figures for our own hearing; we become less sensitive to the tune of 40–65 decibels if our ear-drums and ossicles are removed!

With such sophisticated means of detecting sounds, it should come as no surprise that some fish can actually make sounds. Cyprinoids may do so by the simple expedient of letting off air – blowing raspberries! Other kinds have much more elaborate means of producing a language of squeaks, clicks and grunts. The names of some suggest their vociferous nature – croakers, grunts and drum fish.

The noises made are not always hard to hear. At a range of 2 feet, one toad fish is said to produce a sound of 100 decibels which is equivalent to the noise of an underground train. Others can be heard easily at a distance of 100 feet! A variety of means are adopted by fish to make their sounds. Usually the swim bladder is set vibrating by powerful muscles, or it is set resonating by the rubbing of two bones together, or twanged by the release of a flexible bone.

The purpose of fish noises are varied. In species like catfish that often live in turbid waters, the grunts may help to establish territories and bring the sexes together – rather like the song of birds. Some form of echo-location cannot be ruled out either. Social co-ordination may also be achieved by means of sounds in species that shoal – thus strictly comparable with the call and contact notes of gregarious birds.

5

The Horse

Life on the land: built for speed

Introduction

Those who regularly have a flutter on the Tote, sooner or later may have cause to rejoice in the fleet footedness of the highly 'tuned' modern race-horse. Doubtless, not many winners reflect on the long period of evolutionary research and development that went into their steeds.

Needless to say, the horse is a mammal and not an especially typical one. And yet the merit of the beast lies in the fact that everyone is familiar with it, and that the design shows how the structural problems of propelling a large animal at speed have been overcome.

The horse – or indeed any mammal – is far removed from fish dealt with in the last chapter and yet the land vertebrate's chassis is developed from the fish's frame. Before taking a close look at horses, it may be useful by way of an introduction to follow the fascinating story of the emergence of vertebrates as landlubbers, and the design changes this involved.

The problems of being a landlubber

When fish first started to wriggle their way across the swampy Silurian landscape 400 million years ago, the vertebrates were faced with a set of problems quite unlike any that had been experienced before, and these had to be overcome if they were to make a success of surviving above the water line. On land, rapidly fluctuating temperatures of day and night, sun and shade, and the drying effect of air would tend to parch their damp bodies. Their sense organs – vital windows on the world – would have been adapted to sifting information from a dense watery environment and not from the glaringly transparent and thin air. The changes that ensued to bring them increasing control of body water and the re-organisation of their perception equipment alone are stories in themselves. What concerns us are the mechanical and structural changes, because, for the

first time, the vertebrates began to feel their 'feet' or fins. No longer comfortably supported and cushioned by water, their bodies had to be strengthened to withstand their weight and the fish-mode of locomotion also had to be modified for terrestrial movement.

Why legs?

Those early amphibious fish propped themselves up on their paired fins, which had muscular fleshy bases and acted like primitive limbs, and these also helped to hitch the wriggling animals along. Sliding along on the belly is however a very energetic business, because of the high friction generated between the body and the ground. Nevertheless, providing speed is not essential, this kind of serpentine motion is acceptable; snakes for instance get along very well today without legs, but they cannot move fast; a speedy snake reaches about 7 m.p.h. – a record held by the venomous black mamba.

The evolution of limbs from fins allowed animals to raise their bodies clear of the surface thus avoiding the high energy demands of slithering, and to lever themselves nimbly over the ground. Speeds could therefore be achieved beyond the reach of primitive wrigglers; even in those days a premium was placed upon speed because some amphibians were specialising in being predators and others needed to escape being eaten before their time!

Posture

There is a further trend which should be recognised in the evolution of the mammalian posture. Look at any primitive amphibian or reptile, such as a newt or crocodile, and you see an animal with a rather slovenly body – legs splayed sideways and, while resting, with the trunk slumped onto the ground. Only when the animals move do they raise their bodies and this has to be accomplished by performing

a double-ended press-up. For heavy animals this is a tiring business, as anyone who performs twenty before breakfast in the interests of keeping fit can verify for him or herself – and so as the land vertebrate design became more advanced, the limbs gradually rotated beneath the chest and belly. The trunk thus became structurally levitated for good, and the weight could be effectively borne on the legs without the expenditure of too much muscle power.

Built like a bridge

Once raised aloft, the sheer weight of the head, tail and belly tend to cause the body to bend. It is precisely to prevent this happening in land-living

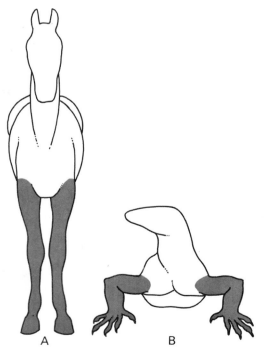

Legs of a mammal placed vertically beneath the body (A) are much better able to support the trunk than those of an amphibian or reptile (B).

vertebrates that the backbone and its associated muscles and ligaments have evolved into a girder for bearing the body. In fish the backbone runs right through the middle and has no supporting rôle to play, but in their amphibious and terrestrial descendants it is relatively more massive and has moved to the top side of the trunk, where it forms a strong supporting arch under which most of the body is slung.

Long ago country lasses found that carrying two pails of milk, yoked over the shoulders, was easier than carrying one, because the weight of the opposing buckets balanced each other. Land animals must also be internally balanced over the fore and hind limbs, and their chassis have been likened to an arched cantilevered bridge, in which the weight of opposing spans are made to balance each other by means of a triangular system of struts, ties, tension and compression members; the triangular layout confers great strength because providing the side lengths remain unchanged a triangle cannot be distorted.

This notion is not so far fetched as appears at first sight because the head and belly, and belly and tail, can be thought of as two pairs of opposing spans balanced over the fore and hind legs respectively. The bridge analogy is made even clearer by considering the structure of the backbone. One of its most noticeable features is a series of dorsally placed projections, or *neural spines* to which the long back muscles are attached. These bony fingers can be related to the struts of a cantilevered bridge. When engineers construct girders, the depth or the length of the struts will, at every point, be proportional to the bending strain on the structure; a cantilevered bridge is therefore deepest over the load-bearing piles where the strain is greatest. So too with the land vertebrate chassis; the strain of bearing the head and belly is usually greatest over the front legs, so the neural spines are longest at that point. In the case of many of the giant dinosaurs, much of the enormous weight of the trunk was counterbalanced by the tail

over the hips, and in these reptiles the neural spines were highest at that point. An interesting sidelight is that, far from dragging their tails behind them, as usually depicted by artists, the bipedal monsters like *Iguanodon* and *Tyrannosaurus* probably strutted around with their tails somewhat cocked to keep the trunk and head high. The kangaroo travelling at full speed is a modern counterpart of a similar design: Immediately one can see the engineering reasons for big tails, 'over-developed' horns, tusks, antlers or other heavy structures. They could play a vital rôle in balancing the body. At a single glance then, it is possible to tell from a skeleton just how the weight of the body is distributed. Many of the larger mammals, including the horse, seem to take most of their weight on the fore legs, and the hind legs are thereby freed for providing most of the power for movement.

Gradually then, the mammalian design emerged as one which was inherently more suitable for life on the land than any of its forerunners. Apart from a superior type of chassis, combining strength without too much weight, a great many other features have led to their supremacy. Warmth, enterprise, ingenuity and care for the young have all contributed to their dominance; so have the waterproofing of the skin, big brains and a good strong bottom jaw equipped with only one set of adult or permanent teeth whose form and function are unsurpassed in the vertebrate kingdom. The selective pressures resulting in the mammalian dental arrangement might have been caused by the need for a high fuel intake to maintain the constant and high body temperature which is also a prerequisite for a high-speed animal like the horse.

Horse power

The horse is not uniquely built for speed. The fact is that the horse displays a whole range of features that have been evolved by medium and large herbivorous

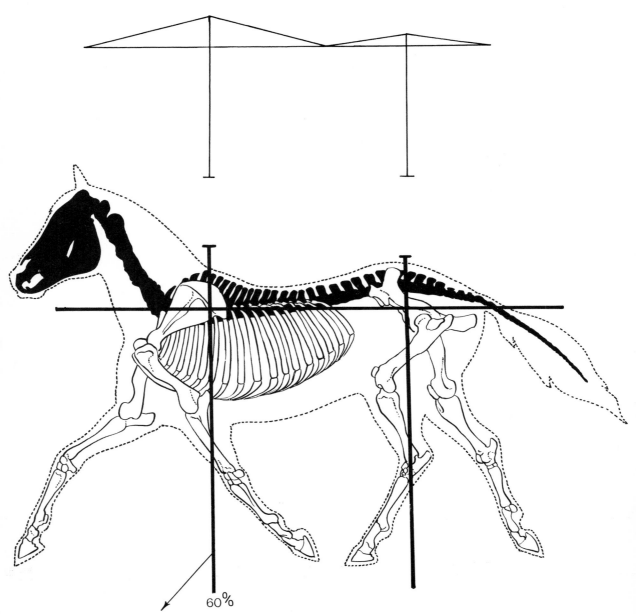

60%

The skeleton of a horse showing how the construction can be compared to a cantilevered bridge. The height of the neural spines give an indication of the bending strain on the backbone at any given point. About 60% of the weight is cantilevered over the front legs.

hoofed mammals – like deer and antelope – in their struggle for survival. Ungulates – as they are collectively called – have devoted their lives to cropping and grazing, an unskilled activity that nevertheless takes a great deal of their time for vast quantities of food has to be consumed and fermented in their very complex gut-processing plants. They have long deep faces for accommodating formidable batteries of grinding teeth which never stop growing to compensate for the wear caused by their rough diet. It is an unaggressive way of making a living, but there is a penalty to pay for thriving on grass and being big and tasty; they must be constantly on the alert from a host of fast and wily meat eaters. Accordingly ungulates are endowed with keen noses, sensitive ears, and big eyes, and to cap it all are usually suspicious and nervous, and take to their hooves at the slightest provocation.

Increase in size is one way of keeping a few paces ahead of the enemy for a larger animal is usually a faster one. Horses have been no sluggards in the race for increased speed. The ancestor of modern horses was a beast the size of a fox terrier *(Eohippus)* and probably lived in the shelter of North American forests. As the line emerged from the protection of thick vegetation, to exploit the coarse grasses of the prehistoric prairies or steppes, speed and stamina became the key to survival. And so a series of increasingly larger and faster horses evolved, culminating in the plains-dwelling zebras, Przewalski's horse, and the crop of asses inhabiting the Afro-Asian steppes.

High-speed legs

What are the requirements of a fast runner? Firstly, the limbs must be long to increase the length of the pace. This has been achieved largely by raising the animal up on the tips of its elongated fingers and toes; in doing so an extra flexible joint has been incorporated into each leg to add more spring to the gallop.

Secondly, the limbs must be capable of moving fast because the speed at which they thrust backwards against the ground will determine the forward velocity of the animal. These two considerations have imposed certain design restrictions on the limbs of large high-speed mammals, because whatever else it may be, a leg is far from ideal in engineering terms for propelling a body. The point is, a limb is a *reciprocating* structure – like a pendulum moving backwards and forwards; once its thrust has been expended on the power stroke, it must be dragged forwards on a recovery stroke using valuable muscle power without doing any useful work. A wheel is the ideal structure for imparting motion, because it rotates in the same direction around the axle and has constant momentum and so energy does not have to be wasted alternately accelerating and stopping the structure. (Forty per cent of the energy of a running man is wasted on constantly reversing the motion of the legs, and this is why we find cycling so much more efficient than running.) However, the wheel is something that is not found in nature possibly because there is no connection between it and the axle. So we and our vertebrate cousins have had to make do with limbs.

Now the momentum (or the kinetic energy) of a limb will depend upon its *mass* and its *velocity* and these are related in the following formula well known to any 5th form physics student:

$$e = \tfrac{1}{2} mv^2$$

With a slow-moving animal (e.g. loris) or a beast which needs measured but extremely powerful movements (e.g. mole) the kinetic energy of their limbs can be ignored as a design consideration. However, for fast animals, it is very much in evidence because the momentum increases with the square of the velocity; *double* the speed and the momentum is increased by *four* times. A fast animal cannot obviously do anything about the speed of its legs, but at least it can keep the weight down. Horses and their kind have eased the power requirements of their swinging legs by reducing the weight of the extremi-

The structure of a horse's rear leg showing how the limb has been elongated below the heel or hock (a). The chief locomotor muscles are situated high up on the leg where a relatively small contraction produces a large movement of the hoof.

the corresponding *metacarpal* bones are fused to form a single strong cannon bone in each leg.

The rear legs are basically the horse's engines, the front ones and the shoulders are admirably constructed for taking the shock of the beast as it pounds over the ground. The impact of an animal weighing upwards of $\frac{1}{2}$ ton, travelling at 25–30 m.p.h. over a six-foot fence is something to be reckoned with, and yet the leading leg must be capable of absorbing the momentum without shattering the leg-bones or rupturing any of the tendons. Horses have four in-built shock absorbers at their front ends. Firstly, the shoulder blades with which the fore limbs articulate are not firmly attached to the rest of the chassis but are suspended by muscles. This arrangement not only allows the shoulders to swing fore and aft, and in doing so increases the length of the horse's front stride, but also allows the front suspension to give a little. The second set of shock absorbers occur at the bottom of each fore limb at the joint immediately

ties, and clustering the power plant or the muscles at the upper end of the legs, which do not move so much. The result can be clearly seen in the hind limbs of any ungulate. The hip girdle is enlarged and nearly vertical, in order to provide anchorage for the powerful limb swinging muscles – as well as those responsible for keeping the belly up. A small contraction of the muscles running between the girdle to the short thigh bones produces a very great swing at the hooves. Furthermore, muscles high up on the leg move the joints around the ankle and 'foot' by means of tendons or 'ham-strings' attached to the hock. This means that the swiftly moving 'foot' can be as light and as slender as possible – a factor that has gradually resulted in the virtual loss of all but one of the digits on each foot; strength for strength, a single bone is lighter than several. The horse runs on a pair of middle fingers and a pair of middle toes, and the tips of these are protected by greatly enlarged 'nails' or hooves. The cloven feet of cattle and their kin correspond to their third and fourth digits on each limb, although

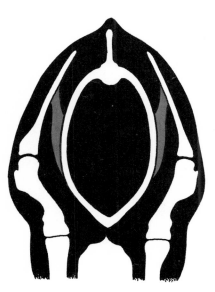

Transverse section through the chest region of a horse showing how the body is suspended from the shoulder by means of muscles.

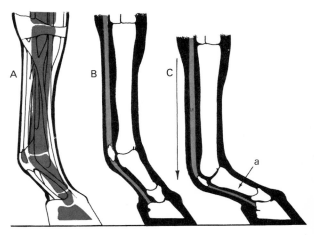

above the hooves (*fetlocks*). This is a flexible joint which is maintained in an almost upright position by powerful tendons running from the 'wrist', behind the fetlock and sloping *pastern*, to a point of insertion underneath the last *phalange* or finger bone which supports the hoof. The arrangement allows some of the shock to be taken up in the elastic tendons. In show jumping, when the impact of hitting the ground is very great, the fetlocks may even touch the ground themselves momentarily. Whether a horse is comfortable to ride may depend to a large extent on the slope and length of its pastern. A hack may have long markedly backward projecting pasterns and is a well-sprung horse whereas a drey weighing a ton or so has upright pasterns with little spring in his feet! – it is only required to lean forward in the harness!

Under normal circumstances, one suspects that horses manage very well, but we have taken the horse's design to its limits with the result that, from time to time, the chassis does fail. One of the most common faults occurs in the fetlock area. When a horse is encouraged to race at between 20 and 30 m.p.h. for three or so miles and to jump a number of fences as well (e.g. in the Grand National) then the animals become tired. When fatigued they start to lose their muscular co-ordination and often land awkwardly after a jump. In these circumstances the

ligaments in the lower leg may become torn and the first phalange (finger) bone below the fetlock may even shatter into a hundred pieces as a result of the nutcracker-like force between the hoof and long *metacarpus* bone. It is a solid little bone which would be difficult to break with a sledge-hammer such is the force of a moving horse! But once broken, the animal must be destroyed for it will never run again.*

The backbone together with the associated muscles and ligaments form a loaded girder, and 60 per cent of the weight is taken on the forelegs. Like most large herbivores, horses have stout hearts and capacious lungs enclosed by well-developed ribs. These greatly add to the rigidity of the body because the breast bone or *sternum* functions as a compression member supplementing the rôle of the spine. The two together form a very deep box girder of tremendous strength. It is not without reason that jockeys sit well forward over the chest where the animal is well constructed for load bearing.

In the horse and other ungulates the lumbar part of the backbone is fairly rigid, and so it is fairly easy to sit on. In other runners and leapers, it is flexible to a degree, and so the up and down flexions of the back

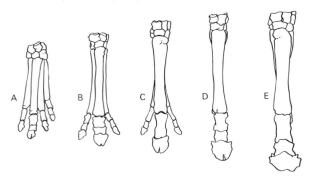

Evolution of the modern horse's hoof (E) from the four-toed Eohippus (A).

*In the course of preparing the television series, two fresh large pastern bones from racehorses were tested under compression. Both broke under a load of 6.9 tons.

Comparison of horse and cheetah at full gallop showing how the swinging shoulder blades allow an increase in the length of the pace. The cheetah, however, in addition has a much more flexible spine than the horse.

increases both the power and the pace of small sprinters. For example, flexing of the spine by well-developed muscles is important in adding spring to the leaping run of rabbits – a fact that could be chewed over when relishing the fine casseroled meat on the back of one of these mammals. The use of the backbone to amplify the movements of the hind legs is perhaps best seen in the small fast carnivores. Whippets and greyhounds, which have been selectively bred for speed, can, by virtue of their mobile lumber spine, bring their back legs well forward in front of their bodies before recoiling them like a spring. Greyhounds can sprint at between 37 and 41 m.p.h., covering 345 yards from a standing start in 17.1 seconds. By comparison, the cheetah, the cat's answer to the greyhound, can probably attain between 50 and 56 m.p.h. when pressed, but like the greyhound, can only keep up top speed for a short time. The fastest on four feet is reported to be the pronghorn antelope from North America which has been reliably reported reaching 61 m.p.h. over 200 yards, but an equally impressive 53 m.p.h. over 1 mile and 35 m.p.h. over 4 miles. Running with the whole body is a very tiring business, and these mammals are generally only good sprinters. Horses are less speedy but have better stamina.

Racehorses can comfortably top 40 m.p.h. for a short burst and can cover a quarter mile in 20.8 seconds, which is a respectable time for some cars driven over this distance from a standing start. Nevertheless, it is salutary to reflect that the car needs both high energy fuels, and a smooth tarmac road, whereas the horse can make do on grass and goes almost anywhere. Despite our ingenuity, no vehicle we have so far built approaches the horse in its versatility. For here is a beast weighing upwards of $\frac{1}{2}$ ton, capable of carrying a payload of two people that can take rough ground in its stride, leap ditches, track up mountains, cross rivers, and clear fences as tall as a man. It is little wonder that empires have been won and ruled from the backs of horses.

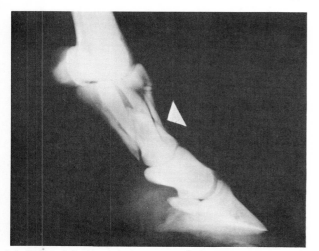

An X-ray of the front leg of a horse (hunter) which has shattered its greater pastern bone.

6

The Elephant: a modern giant

Introduction

There is nothing quite like an elephant. Its shape is unique enough, but the sheer size of the animal is its most impressive feature. No other surviving land-based creature can match a bull African elephant's 10 tons or more; in practical terms this means it would take 15 mini-cars to balance one of these beasts on a set of scales. By comparison, the white or black rhinoceros and hippopotamus are both in a different class, although two of these species can reach a respectable 4 tons each. Nevertheless, the African elephant would appear puny beside some of the dinosaurs that used to plod around this planet. These reptilian heavyweights are perhaps the most intriguing of animal giants; unfortunately they have been extinct for 70 or more million years, and so the elephant will have to do as the best living example of how nature has coped with bulk.

The problem of being big

Big beasts are not simply scaled-up versions of smaller models. This principle is rather difficult to illustrate using a series of elephant types as examples because, apart from the hyrax and sea cows, the Indian and African elephants have no very close living relations. The rodents do, however, span quite a size range, from the tiny pigmy mouse of Africa, which can practically sit on an old penny, to the South American capybara, which would need an armchair. The former is a lightly-built mammal with legs like match-sticks, whereas the latter is robust and is supported by stocky limbs powered by relatively large muscles. If the capybara was scaled down to the size of the pigmy mouse without changing any of its proportions, one suspects that the result would be a rather cumbersome little animal with supports and muscles way and beyond its needs. However, what is more important, if the mini-mouse were scaled up to the capybara's dimensions, then it would probably collapse under its own weight and certainly fracture all of its legs as soon as it had to run for its life.

The point is this: the weight of an animal increases proportionately much more than its dimensions. Take two beasts, one of which is *twice* the dimensions of the other: the larger will be to the order of *three* times the weight of the smaller. In other words, the weight varies with the cube of the linear dimensions.

So in an evolutionary progression of increasing size, the result of which has culminated in fine giants like elephants, the animals become relatively heavy. Now an engineer faced with such an overweight problem, as when 'stretching' an aeroplane, can use lighter and perhaps stronger materials to compensate. Unfortunately, big beasts are built of the same stuff as their smaller ancestors, and so relatively more bone and muscle must be incorporated into their design in order to prevent them falling apart, or providing them with acceptable safety margins.

The strength of a bone or the power of a muscle is proportional to its cross-sectional area. If an animal doubled its size without altering its proportions, normally it would double the strength of its leg bones and the power of its muscles, *but* it would be trebling its weight. To keep itself in good mechanical trim it would have to increase the size of its legs and the associated musculature by a factor of three. It is therefore easy to observe that big mammals have more powerful and stouter legs than smaller ones.

The elephant's legs are one of its most noticeable characteristics. They are sturdy, and unshapely in the extreme, but nicely designed to take the weight of the animal and to propel it at a creditable 30 m.p.h. if necessary. The big bones are massive and while held vertically beneath the body have no difficulty in supporting 10 tons or so between them. Elephants, however, are very mobile, and when on the move or performing balancing acts in circuses may have two legs off the ground at any one moment, so clearly there must be a considerable safety margin in the strength of the bones to cope with much higher

African elephants.

tensile forces. Nevertheless, one suspects that stumbling might have much direr consequences for an elephant than for a mouse!

A further characteristic of the elephant's adaptation to supporting his weight is his comical shuffle. When they move they do so without any spring in their heels. The reason is that although their wrists and ankles are raised in relation to their fingers and toes, they are permanently so and are encased in muscle and connective tissue. The feet therefore may act as a kind of shock absorber, but no extra push can be produced by raising the heels, so the elephant must walk in a permanently 'flat-footed' way. Despite their size and power, they have no chance of

jumping and a miserable 7-foot-wide trench can imprison these animals as effectively as reinforced concrete walls.

The strain on the bones can also be kept to a minimum by keeping the legs as near to the vertical as possible, so the elephant's lumbering is yet another adaptation to its mass.

A digression on dinosaurs

The discussion in the previous section does raise the question as to whether there is any theoretical limit to the size of an animal. Even today, the African elephant is not the largest beast in the world; the

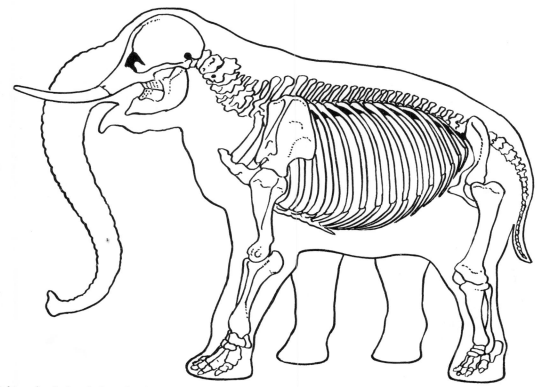

The skeleton of an Indian elephant showing the massive bone structure of the legs.

An African elephant, white rhinoceros, and the extinct Brontosaurus *to scale. The latter is seen from a three-quarters side view, exaggerating the size of the head, which was in fact much smaller than that of a rhinoceros.*

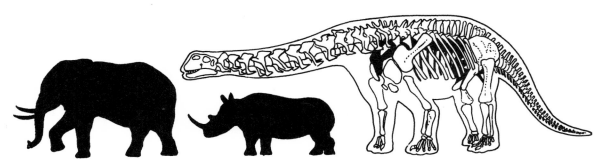

blue whale ranging up to 150 tons may conceivably be the animal goliath of all times. Moreover it is a marine mammal, and while immersed in water has no weight problem at all; however, stranded whales are likely to be crushed by their own mass and overheat into the bargain. So far as terrestrial animals are concerned, modern elephants hardly stand comparison with some of the dinosaurs. A full grown *Brachiosaurus* could cover one and a half cricket-pitches, being 90 feet from its ludicrously small head to the tip of its tail, and probably topped the scales at between 50 and 80 tons. There is, however, some dispute as to whether this monster reptile was utterly terrestrial or whether it was amphibious, in which case much of the weight would have been taken off its feet. Other sauropod dinosaurs were undoubtedly land-based. *Apatosaurus* weighed 30 tons, and calculations made by Professor R. McNeill Alexander of the University of Leeds show convincingly that, providing the animal did not take too long a step, the forces exerted on the long bones would not approach too closely their theoretical breaking strain of 20 tons. Terrestrial animals larger than *Apatosaurus* are doubtless possible, but these would become less agile and would tend to move their column-like limbs in a stiff-legged shuffle.

Reference has already been made in the last chapter to the analogy between the land vertebrate frame and an arched cantilevered bridge. In the case of quadrupedal giants this is particularly marked. Their spines are placed high up and arched (an arch is a very strong structure and will bear enormous loads, as early masons and builders found). Dinosaurs like the 85-foot *Diplodocus* sported a long neck and tail, and these were cantilevered out from the body, and suspended by tendons which ran through the forked neural spines.

And then there is the matter of fuel supply; animals weighing tens of tons would require fantastic supplies of exceedingly rich food; the sea can provide this, the question is, can the land? Well-fuelled bodies produce heat and it has been suggested that, with increasing bulk, land-based animals could possibly run into problems of dissipating their excess body heat, particularly if they were warm blooded and lived in tropical conditions away from cool wallows. Heat is most easily lost from the surface of the skin, and with increasing dimensions the amount of skin area becomes relatively less. An African bush elephant has enormous ears which are intensively vascular, and, when held away from the body, they increase by one-third the area over which heat can be convected or radiated away. How the large dinosaurs managed is a matter for conjecture. However in *practical* terms, with the possible exception of *Brachiosaurus*, the 30-ton *Apatosaurus* seems to represent the upper size limit in purely land-based animals.

How the elephant got its trunk

Who would believe in elephants unless they had seen them? With their trunks and tusks, they are perhaps

the most improbable of beasts alive today. Early
artists variously ascribed to elephants – doubtless
based upon too little knowledge or ignorance – in-
built trumpets or hose pipes. So far as their structure is
concerned, one could say that the evolution of the
family is synonymous with the development of that
wonderfully flexible organ, the trunk, and that this is
not totally unconnected with their large size.

The elephant story starts not with a giant, but with
a lilliput, going by the name of *Moeritherium*. It was
rather like a pigmy hippopotamus standing 2 feet at
the shoulder and spent its time lazing around in
swamps 60 million years ago in the Eocene period. A
pair of incisor teeth in the upper jaw had already
formed into small tusks. Without implying that
Moeritherium was the ancestor of all subsequent
elephants and their kin, one can say that a beast not
unlike this little Eocene mammal was at the root of
the elephant family tree. From here, the trend was
towards rapidly increasing size and elongation of the
face – and lower jaw. This trend resulted in the
evolution of animals like *Phiomia* with a long lower
jaw, and a matching upper one, possibly terminated
by a very mobile upper lip or snout that would be
invaluable in helping food into the mouth rather like
the tapirs have today. At this stage the elephant
family included some fine species, known collectively
as the spoon or shovel tuskers. *Platybelodon* had an
enormous lower jaw terminating in two spade-like
incisor teeth. The whole mouth was probably used as
a scoop for taking in plants from the bottom of
lagoons. They were all specialised animals that were
offshoots from the mainstream of elephant progression.

The development of the 'long muzzle' may have
been a direct consequence of increasing size. As
the animals became taller they probably found it
difficult getting their mouth to the ground for feeding.
The hippopotamus has solved this problem by having
short legs, and the wart-hog may go down on its
wrists to graze, while another method would have been

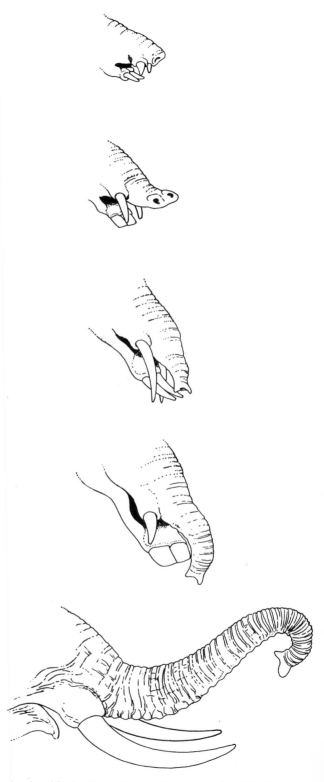

*A series of Proboscid muzzles showing the stages by which the trunk
could have evolved (refer to text for detailed explanation).*

to evolve long necks, but, since the days of *Moeritherium*, elephants have never had much of a neck. It is therefore conceivable that those individuals that retained elongated faces and muzzles were at an advantage as the animals became tall, because they could still reach the ground and crop their food low down. What happened next was a phenomenon that students of evolution find over and over again – that a structure or organ that served one particular purpose takes off and undergoes a rapid series of changes so that it comes to serve another quite different function. A point was clearly reached in those early long-faced elephants, where the small mobile snout became more useful gathering food than merely acting as an adjunct to the mouth. The forces of natural selection then operated in such a way that the lower jaw began to fall back, followed by the bony support of the face, and the fleshy muscular snout became greatly elongated to form the trunk. Together with the tusks which are themselves glorified upper incisor teeth, it came to form a collecting and manipulating organ as sophisticated as we could hope for from a collection of nose muscles!

Despite its elevation, and no neck worth speaking of, an elephant can efficiently crop grass or reach perhaps 18 or more feet into a tree to pull off branches containing succulent leaves. With one sniff it can draw up to $1\frac{1}{2}$ gallons of water and squirt it down its throat; and when water is in short supply, the trunk can be used to dig into dry river-beds to reach the water-table. Equally important, dry earth can be blown onto almost any part of the body as part of the elephant's skin care, and different trunk postures signal the elephant's mood in a kind of semaphore language. The tip of the trunk is drawn out into 2 or 3 lobes – depending upon the species – and these are not only highly sensitive to touch but are capable of handling small objects almost as deftly as our own finger and thumb. Although the trunk is capable of gentle movements, it can also wield great power. Quite large pieces of wood can be broken by placing

them across the tusks and then bearing down with the trunk.

As a design, it is possible to say that the elephant form has been a remarkably successful one up until today. There have been animals which were recognisably elephants for the past 26 million years; all were big, sported trunks, and were adorned with tusks of various shapes and sizes. *Dinotherium* had tusks curved back on themselves; what purpose they served and how they were used is anyone's guess. Then there were the impressive forest-living mastodons, and the mammoths, one of which was a shaggy arctic beast, the remains of which are still being released from the Siberian permafrost. Some of these had tusks $16\frac{1}{2}$ feet long and the way they are worn at the tips suggest they might have been employed as snow ploughs for exposing the coarse tundra vegetation. The African elephant's tusks can be used both for digging and as weapons. The world record weight for a *single* tusk is 106 kilogrammes for a bull, and 25 kilogrammes for a cow. Theoretically a bull African elephant could support a pair of tusks over 18 feet long and weighing 3 cwt when they were 60 years old; wear and breakages means that this is probably never achieved. Even so, with such a weight of ivory on their mind, so to speak, it is not surprising to find that the elephant's skull is a bulky affair but at the same time extraordinarily light, for much of the structure is filled with air sinuses reinforced by a lattice-work of bone. In some ways this is similar to the honey-combed structures used in our own building materials which combine strength with lightness.

Teeth to last only a lifetime

Although the elongated incisor teeth are the most conspicuous, those hidden inside the elephant's mouth are by far the most important for its survival for their chopping and grinding action break down the fibres and plant flesh in readiness for the digestive

The elephant's family tree. (A) African elephant. (B) Mastodon. *(C)* Trilophodon. *(D)* Dinotherium. *(E)* Moeritherium.

process in the stomach and guts. The teeth themselves are nicely adapted for grinding, being composed of a series of transverse plates of bone-like *dentine* and *enamel*, set in a matrix of *cement*. These substances wear differentially and form a ribbed grinding surface not unlike a miller's stone.

Perhaps the most interesting feature of the elephant's denture is that the teeth are budded off from the growing area at the back of the jaw, one at a time, and as each is worn down a new one gradually moves forward to replace it. In keeping with the animal's need for more food as it grows, successive teeth are larger; a 9-month-old calf has about 20 cm² of teeth in wear, whereas a 48-year-old bull has 650 cm² of grinding surface with which to masticate his meals.

Throughout life, an elephant will rely upon only 24 teeth (6 in each half-jaw) and when the last molars are worn down, feeding becomes inefficient and the animal finally weakens and dies. So, if disease, drought or some other misadventure does not overtake an elephant before this time, he will finally succumb around the age of 60 from mechanical senescence.

Have giants a future?

It could be argued that despite their size and strength, giants are peculiarly vulnerable creatures. Certainly they never came off particularly well in fairy stories! And yet some animal giants have proved remarkably resilient and have had long runs of evolutionary success. The dinosaurs, although usually held up as examples of animals that failed to make the grade to the present day, were in fact an extremely long-lived group and dominated the land fauna for 70 million years; this makes our own million or so seem a trifle insignificant! Many giants have come and gone, and the evidence indicates that even the elephants have had their hey-day. Out of 350 or more Proboscids known to zoologists, only two remain – the African

and Indian species – and, as we shall see, the long-term fate of these is in considerable jeopardy.

Why do selective processes time and time again seem to favour a trend towards increasing size, and why do the resulting giants tend not to survive for too long (on a geological time scale)? The usual reasons given for the first part of the question runs as follows. An animal which is larger than the other members of its kind is stronger and more powerful in competition for food and mates; it can outrun enemies more effectively, and, relative to its bulk, needs less food. In engineering terms it is a more efficient machine. A large warm-blooded animal can also survive more easily in cool climates because it can conserve its heat better; thus a mammoth could keep warm on the open arctic tundras where a mouse would quickly perish. Nevertheless, the advantages of being an outsize animal have to be paid for. Populations of big beasts require plenty of space and prodigious quantities of food. A tiny shrew weighing a few grams may consume twice its body weight in worms, grubs, and insects each day; an elephant only needs between 4 and 6 per cent of his mass, but this still represents a daily intake of $1\frac{1}{2}$ cwt of vegetation; for an elephant's life is one long meal – 16 hours each day is needed to refuel his body!

The environment must be fairly lush to provide food for a viable population of giants. Not unrelated to this requirement is the fact that big mammals have relatively low populations when compared with those of small species. They also breed more slowly. Cow elephants reach puberty between 11 and 20 years of age, and over the course of 45 or 50 years of reproducing may give birth to 11 calves. One suspects that because of the low rate of turnover, and the comparatively small populations, giant animals cannot respond very quickly to dramatic changes in the environment. It is therefore possible that big beasts evolve in very slowly changing circumstances of climate, vegetation, and so on, and thrive best when conditions are reasonably stable. (Note the whales which live in perhaps the most stable of all environments – the sea.) There is little doubt that gross size is, in a sense, a specialised feature, but like all specialisations, it can prove to be a blue-print for rapid extinction when the conditions to which the animal is adapted start to alter. When this happens, it is better to be a generalised mouse rather than an elephant.

Even a cursory glance over the fossil record will show that a large number of giant mammals have been lost for good over the past million years, many during historical times. In general, this planet has suffered some very severe climatic upheavals over the course of this period, whereby the polar ice caps have alternately expanded and contracted a number of times. Many of the fatalities can be attributed to what has been called, the Ice Age over-kill. Many however cannot. When the glaciers retreated between 12,000 and 30,000 years ago, most continents sported fine mammal communities comparable to those in East Africa today. America had its giant sloths, super camels, majestic forest mastodons and many others. In Europe, there were mammoths, giant cave bears, mighty aurochs, woolly rhinos and Irish elk. These big beasts were the first to vanish, and the evidence is that man the hunter – or more appropriately the undertaker – was responsible in part if not in full for their ignominious fate. The lesson is simple; with such low populations and slow potential for making good high mortality, big beasts are easily wiped out by an efficient predator, or forced out of a living by the commandeering of their homelands for farming of one sort or another.

Today we are left with only a handful of what could be called respectable terrestrial giants; two elephants, three heavyweight rhinos, a hippopotamus, and, stretching a point, the giraffe. The long-term survival of none of these is good. The African elephant is the most numerous; there are $\frac{1}{2}$ million left of which 150,000 are inside National Parks where many are causing management problems on a gigantean scale. Confined in these comparatively small areas, with

their numbers unchecked by poaching, they are changing the bush and woodland into grassland by destroying the trees. When an elephant gets really hungry he just leans on a tree and destroys in seconds what's taken one or two hundred years to grow. White rhinos in South Africa – a species only just saved from extinction earlier this century and now numbering 2,000 or so – are over-grazing the grass upon which they live in the Umfolozi National Park, Natal.

The sad fact is that, we cannot afford big beasts wandering around at liberty. Their future lies in strictly managed parklands or zoos. But for how long? Based upon the record of the past 10,000 years, a bookmaker would give very small odds indeed on any of them reaching the Third Millennium, let alone the Twelfth! At least we can console ourselves with the thought that long after *Homo sapiens* has disappeared into oblivion, the forces of natural selection may begin to operate on the survivors, and who knows, in the distant future, mighty mice or super shrews may replace the giants that will have vanished with us!

7

Return to the sea

Whales, dolphins and porpoises

Introduction

Thanks to recent developments in the animal enter-tainment business, there can be few people who have not been able to observe small whales at first hand, for upwards of fifty bottle-nosed dolphins are delight-ing audiences regularly by leaping through hoops, playing nose-ball and having their teeth brushed in fifteen or so British dolphinaria. Visitors to these places might well reflect on the skill of the performers and compare them with the banal patter delivered by the impresarios and showmen. However, there is one piece of zoological information that is common to all of the commentaries, whether good or poor, namely that the animals being put through their paces are, despite their appearance, not fish, but warm-blooded, air-breathing mammals – like us. To the non-zoologist, it is probably their boisterous spirits and playful – almost mischievous – behaviour that dis-tinguishes these performing whales from fish, and yet from their shape one could easily slip into accept-ing them as such; a dolphin is nevertheless more closely related to a wolf than to a shark!

At a glance, members of the whale family (or *Cetacea*) have no gills and gill slits, and have instead one or two nostrils on top of their head; their skin is not covered with scales, indeed, some sport sensory hairs or bristles around the mouth; they bear their young alive (so do many fish) but these are suckled in the typically mammalian manner. Now this com-parison between fish and cetaceans may seem rather laboured, and yet it is precisely their differences and similarities to fish which make whales and their relatives so fascinating. For here is a group which, like all mammals, have themselves ultimately evolved from fish-like animals but have returned metaphori-cally to the happy hunting-grounds of their ancestors. In doing so, they have evolved a fish-like form, but this was moulded and developed from a frame adapted to life on the land, for the immediate ancestors of whales were probably carnivorous mam-mals (*Creodonts*) not unlike the modern dogs. The structure of these marine beasts then shows how the terrestrial mammalian frame and organisation has evolved to cope with a fish-like way of life.

Shape

Any large animal moving through a dense fluid, like water, comes up against similar problems of friction which manifests itself in the form of drag. These forces tend to slow the animal down, and a great deal of energy is needed to overcome them. Accordingly, marine animals that are divorced completely from the land, and do not have to compromise their structures to support their weight when resting on the shore, tend to evolve streamlined shapes whereby turbu-lence is kept to a minimum and muscle power can be correspondingly economised. A dolphin's body in-creases in girth to a maximum width well over half way back between the snout and tail; streamlined hulls of this shape cut down eddy formation to such an

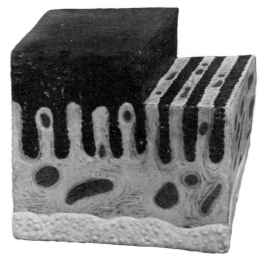

Model of the skin of a dolphin showing the blubber and dermal ridges which 'key' the skin on to the underlying tissues.

The swimming movements of a Common dolphin. The chief power stroke is the upward movement of the trunk. Arrows indicate the flow of water past the body.

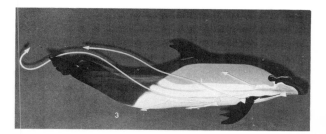

extent that the drag may be reduced to nearly one-half that of a comparable one producing turbulent flow. The effectiveness of the dolphin's streamlining is shown by an observation made of these mammals swimming at night among phosphorescent plankton. It was noticed that turbulence tended to stimulate bioluminescence; when observed from a ship, seals, swimming among the dolphins, left a bright luminescent wake whereas the small, fast-swimming whales left a trail which was far less conspicuous.

The power required by swimming dolphins and whales is a subject that has interested zoologists for a long time, particularly since it was discovered that, despite its streamlined shape, a rigid model of a dolphin needed far more energy to propel it through water at high speeds than a cetacean of similar size could theoretically muster from its muscles. We know that most cetaceans are capable of at least 20 knots, although the enormous blue whale may only be able to keep this up for about ten minutes or so; a big whale would still need to develop over 500 h.p. Since a dolphin's heart is no larger than our own, and, weight for weight it cannot refuel the muscles at a faster rate, it seems improbable that whale muscle is inherently stronger than our own.

The solution to the power puzzle may well lie in the cetacean skin which acts as a pressure sensitive diaphragm. Resting on a great thickness of blubber, which is itself not rigidly attached to the underlying tissues, the skin probably oscillates freely as the animal moves through the water, and the effect of this may be to dampen the turbulence in the boundary layer. If a model is covered with dolphin skin, the drag registered while it is being pulled through water at different speeds is significantly reduced.

A further feature of the cetacean skin is worth considering. As the mammals surge through the sea, the friction created by the flow of water past the skin will tend to drag it rearwards. To prevent this sloughing effect, the outer part of the skin is 'keyed' onto the deeper layers by ridges. These are rather similar to

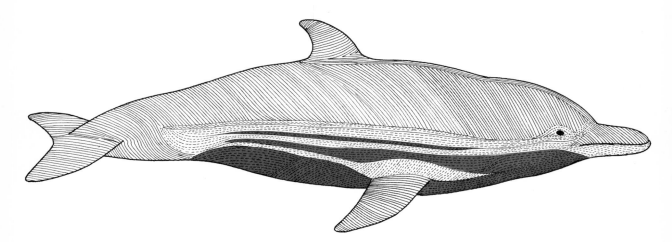

those on the skin of our own hands, which forms a rough surface well adapted for gripping. Now the 'internal' fingerprints on the cetacean skin are not haphazard; far from it, because they tend to run parallel to the overall water flow over the body, and are therefore oriented in the best way to counteract the sheering forces generated by the friction. In more ways than one, then, the dolphin's supremacy in swimming is skin deep!

The cetacean shape does differ fundamentally from the fish's form. The most obvious point of departure is in the orientation of the tail flukes. In fish, the caudal fin is *vertical*, whereas the whales sport their flukes in a *horizontal* plane, and they are not supported by a series of bony rays. The caudal fin of fish and the whale's flukes have no relationship to each other at all; they have independent evolutionary histories, although they both perform the same functions – that of producing thrust. Their different orientation reflects their separate histories. As stated in a previous chapter, fish retain segmented muscle blocks placed in either side of their backbone, and by their rhythmic contractions, the trunk is made to flex from side to side. However, the land living mammals that produced the cetaceans had advanced their design to such an extent that the back muscles had lost their segmented arrangement and had become arrayed in

sheets and strips anchored at one end to the vertebral column and at the other attached either to the shoulder and hip girdles, or to the head and limbs. Related to rapid transit over the ground, the backbone tended to evolve a degree of flexibility in a vertical plane, thereby leaving the old fish-reptilian method of wriggling behind them (see chapter on the Horse). When whales took to the water, some form of wriggling again became the order of the day, but this had to be developed from existing muscles and modes of behaviour. To those who have watched seals and sea lions, or even sea otters hitching their hind parts along, it should come as no surprise that dorso-ventral flexing of the trunk was adopted by the whales for propulsion, and, accordingly, natural selection resulted in the evolution of horizontal flukes from outgrowths of the body wall.

With the adoption of a totally water-borne way of life, the backbone and limbs became redundant as weight and stress bearers. The former reverted to its original rôle as compression strut to resist any shortening of the trunk when the dorsal and ventral back muscles were operating. In whales, the trunk vertebrae are relatively simple cylinders of bone with dorsally-arranged processes for the attachment of muscles; the projections which tend to lock the bones together for strength and rigidity as found in many

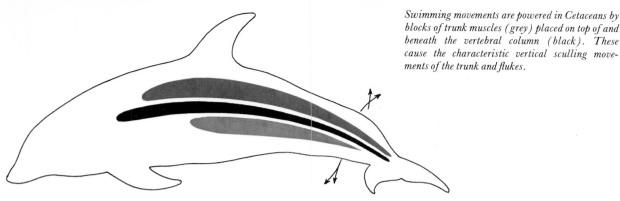

Swimming movements are powered in Cetaceans by blocks of trunk muscles (grey) placed on top of and beneath the vertebral column (black). These cause the characteristic vertical sculling movements of the trunk and flukes.

terrestrial beasts are usually reduced.

Most fish have two sets of paired fins corresponding to the front and hind limbs; cetaceans have only one pair to aid stability. These are derived from the front legs, although the 'fingers' may be elongated and consist of more bones to support the 'blade' effectively. The hind limbs and hip girdle have sunk almost without trace. Only two tiny bone rods buried well within the body wall remain, and these have probably been retained because the penis muscles are attached to them!

Stability is further increased by the evolution of a small dorsal fin, although like the flukes, this is an outgrowth of the body wall, and is not stiffened by bones. In the fast-swimming and highly manoeuvrable killer whales it is particularly well developed.

Other features are worth mentioning which are a direct result of the whale's adaptation to full-time swimming. The testes of the male are housed inside the body, presumably as an aid to streamlining. The eyes too are modified for seeing in water; like fish, the transparent 'window' or cornea is flattened, and the lens has assumed almost entire responsibility for refracting the light on to sensitive retina. Accordingly, it is almost spherical. They have little need to see upwards and so their eyes tend to be directed sideways and downwards. Whales have also little need for tear glands and ducts; these have thus vanished.

How to keep warm

The sea is no place for a warm-blooded mammal unless it is well protected against the cold. Whales have long since lost their insulating fur coats, presumably in the interests of cutting down friction. They have replaced their fur with fat, for every cetacean, seal and sirenian for that matter, is enwrapped in a great roll of blubber. Some of the smaller species may consist of nearly one half of their weight of fat. The insulating properties will depend upon the surface area of the whale relative to its bulk, and how thick the layer of blubber is beneath the skin. Small dolphins have a layer about 2 centimetres thick, and this is barely adequate for preventing body heat sapping away. The notion has been expressed that these small species must keep moving in order to make good their heat loss. Indeed, it is doubtful whether a cetacean could survive if it were smaller than the La Plata dolphin, which is $5\frac{1}{2}$ feet long and weighs 88 pounds. This species is only marginally larger than the 85-pound sea otter, which ranks as the smallest marine mammal in the World.

Many species of cetaceans migrate into warm waters to give birth, thus possibly alleviating heat loss problems for their small calves.

The bulkier species become better off in their ability to keep warm; they have relatively less surface area in

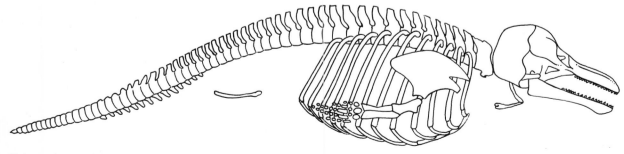

Skeleton of a porpoise. Note how the pelvic girdle has all but vanished and how the hand-like bone structure supports the pectoral 'fins'. The flukes have no skeletal support at all.

contact with the water and have thicker layers of fat. A rorqual needs 17 centimetres thickness of fat to keep warm while resting; some species have a blubber coat 60 centimetres thick – a coat which has been and still is greatly prized by us, and which has contributed to the demise of the larger whales.

Apart from its insulating properties, the presence of so much blubber is doubtless important for buoyancy, as well as acting as a vast food store, and a potential store of fresh water, for oxidised fat yields water.

Breathing and diving

If cetaceans could be taken back to the drawing-board, it is interesting to speculate on whether any changes could be made to their oxygen life support system. Bearing out the principle that evolution does not operate in reverse, whales have not re-evolved gills but have been lumbered with the lungs of ter-restrial ancestors, and all that goes with them. Despite their streamlining, and ability to hunt, feed, and mate under water, these fine beasts are as surely tied to the surface as we are when we sally forth to explore the sea-bed.

The respiratory tract and the process of respiration have, however, been modified to allow them to spend as little time as possible at the surface. Most obvious is the position of the nostrils, which have moved from the tip of the snout to a prominent position on top of the head. They can thus breathe without too much of their head breaking the surface, and producing a bow wave which is extravagant on power (movement on the surface is always wasteful because of the turbu-lence it creates – thus whales should be compared with submarines rather than ships). In some species the area around the blow-holes is sculptured in such a way that water is deflected from the orifices when the animal is taking air. A large whale's lungs may con-tain 2,000 litres of air at sea level.

Of course, whales do not spend much of their time ventilating their lungs at the surface; many dive considerable distances searching for squid and fish; a 47 foot long sperm whale was pulled to the surface entangled in a submarine cable which was lying in 620 fathoms (3,720 feet) of water. At this depth the whale would have been subjected to a pressure of 1,680 pounds per square inch. Even if this record is exceptional, it is known that whales regularly sound to between 20 and 250 fathoms. For such journeys they must be able to take down sufficient oxygen, hold their breath for long periods, and have lungs adapted to withstanding crushing water pressure. For their size, whales have not particularly capacious lungs. On the contrary. Their lungs occupy between 1 and 3 per cent of their body volume compared with 6 to 7 per cent for our own. They can, however, thoroughly ventilate them when breathing. In one breath they can expel 90 per cent of their lung contents, thus getting rid of the stale air, and recharging them with fresh gas. (We usually change only 10 per cent of our air with each breath.) The efficient breathing is made possible by the evolution of a very oblique diaphragm and mobile ribs which are able to squeeze the gaseous

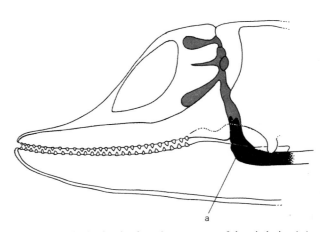

Head of a dolphin showing how the upper part of the wind-pipe (a) can plug into the nasal passage. This allows the animal to swallow prey under the surface without taking water into the lungs.

contents out of the elongated lungs. Apart from storing oxygen during a dive, a lot is stored in the blood and the muscles themselves, which are rich in *myoglobin* (a pigment rather like *haemoglobin* in the blood which has an affinity for oxygen). It has been calculated that a large fin whale can take down a total of 3,550 litres of oxygen (400 in the lungs, 1,400 in the blood, and 1,750 in the muscles) and since it uses about 200 litres each minute, the supply should suffice for a dive of 18 minutes. In fact, they can dive for much longer if necessary. It is very probable that certain of their body processes are slowed down automatically when they submerge; for example the heart beat is reduced when they dive and this must save on the oxygen consumption. It is also likely that a proportion of their energy is obtained without utilising oxygen, so that this valuable commodity can be retained for vital organs such as the brain. We do the same when sprinting, but our own body can only build up a small 'oxygen debt', because we are physiologically very sensitive to the accumulation of lactic acid and carbon dioxide – formed by the anaerobic breakdown of sugar fuels. In ourselves, it is the accumulation of carbon dioxide in the blood rather than oxygen deficit which brings about the lung-bursting sensation when we try holding our breath under water. Whales may be able to withstand very high concentrations of carbon dioxide and lactic acid, and they can ventilate the lungs sufficiently at the next visit to the surface to bring the concentration of both down to a very low level before sounding.

The other adaptation that whales show is their freedom from unpleasant conditions like caisson sickness, commonly known as 'the bends'. This illness is caused by the liberation of small bubbles of nitrogen in the blood stream and tissues in divers who have been breathing compressed air; the 'bends' occur when they return to the surface with too much haste and the pressure is released rather like in a bottle of pop with the top suddenly taken off. Whales largely avoid this complaint because they only take down one lung full of air at sea-level pressure so they physically do not contain a great quantity of nitrogen under pressure. In this respect, they should be compared with a snorkeller rather than a diver. However, as a safeguard, the cetacean lungs are constructed in such a way that as the water pressure on their body begins to crush them, the air is forced into the thicker pipes of the bronchial tree where it is less easy for the gases to be absorbed into the blood. It is also maintained that special sinuses at the base of the skull may help to absorb the nitrogen in oil.

Singing and hearing in the sea

One of the more tiresome operations that a snorkel-diver has to perform as he or she descends into the depths is that of equalising pressure on both sides of the ear-drums. Failure to do this by swallowing hard or blowing hard through the nose pinched by finger and thumb will likely produce intense pain at the best, or a ruptured ear-drum at the worst. The reason for this is that the middle section of the ear is an air-filled compartment containing the three minute ear bones which convey the vibrations from the flexible ear drum to the fluid filled cavities of the inner ear where the decoding of the sounds starts. In order to work properly, the pressure in this enclosed cavity must be similar to the pressure acting on the ear-drum or else it would become too taut to resonate properly – this is why your hearing is temporarily impaired when you are subjected to sudden changes in pressure (e.g. going up in an aeroplane). There is, however, a tube leading from the back of the throat to the middle ear *(Eustachian tube)* which allows air to escape from, or enter the space so as to balance the internal and external pressures.

As can be imagined, the problem faced by whales is monumental in comparison to our own, because of the tremendous pressure changes they normally encounter when diving. Accordingly their ears are considerably modified to meet demands of deep-sea

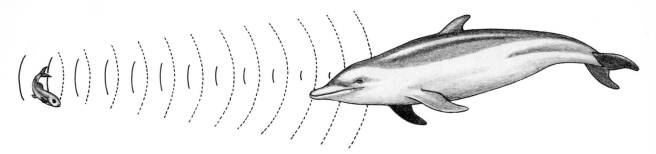

Very high-frequency clicks produced in the Cetacean's nasal passage are focused into a concentrated beam and used for echo-locating.

diving. For reasons of streamlining they have no external ear paraphernalia; indeed, they have little or no external signs of an ear opening. They do not need one because they are practically transparent to sounds underwater, because sound waves are transmitted without difficulty through the tissues. However, they have been endowed with an air-filled inner ear, the volume of which must be maintained constant at all depths. Now the head of a whale has several air sacs connected to the Eustachian tube which is, in turn, open to the respiratory tract. As the whale sounds, pressure causes the air in these sacs to become compressed, and as one after another of the sacs collapse the air in them is forced under pressure into the middle-ear cavity which is thereby kept inflated. The reverse process happens as the animal surfaces.

We can judge the direction from which sounds come partly by assessing (unconsciously) the difference in time taken by the sound front reaching first one ear, then the other, and by the sound shadow caused by our dense head. Sound travels so fast underwater that the former method is impracticable to whales, and the transparency of the body to sounds underwater makes for very weak shadows. The structure of the whale's skull is designed partially to insulate each ear, making each one receive its sounds chiefly from one direction. This is achieved by the strategic arrangement of oil and mucus foam with poor sound transmitting properties, and the presence of *tympanic bulla* bones, which are much denser than the remaining parts of the skeleton.

With such finely developed ears, sound obviously plays an important rôle in the life of cetaceans. For a long time it had been known that the smaller dolphins and porpoises had a repertoire of clicks and whistles, presumably made by squeezing air past valves in the head. Some of these noises undoubtedly function as sonar or asdic, particularly the clicks that range up to 100 KHz (20–25 KHz is the limit to our own hearing). The construction of the dolphin's head allows the

intense sounds to be projected in a highly directional beam, and cause anything in the path of the animal to 'ring' or throw back an echo. Fish with swim-bladders or other whales with their internal air cavities would be particularly effective in reflecting sounds just in the ultrasonic range; fish-eating cetaceans that could echo-locate their prey would therefore be at a much superior advantage when compared with ones that hunted only by sight.

Recent research has revealed that even the large whales have creditable voices – indeed the 'songs' of the humpback whale have been riding high in the record charts in the United States; they have even been orchestrated! The baleen whales utter ultrasonic pulses of the type that biologists use to locate dense shoals of planktonic animals, so it is quite conceivable that the giant blue whales are able to home in on their crustacean prey. Although perfectly adequate for the job, scientists studying the vocalisations of blue whales found that the pitch of the sonar was higher than the theoretical optimum for the size of the prey. However, they concluded that the whales sonar pitch was probably a compromise between the best frequency for locating planktonic shrimps and the need for a good range, because low-pitched sounds carry much farther than high-pitched ones. In this context, it is therefore interesting to recall that the sounds used for communication are low-pitched whistles and moans with tremendous carrying power. Some may carry for hundreds of miles.

Water babies!

Being born at sea could be a dangerous business for an air breathing mammal for death by drowning is always a serious possibility. The fact is that whales have specialised behaviour patterns which help their new-born offspring survive. The flukes are usually first to emerge, and as soon as the head is free the mother nudges the calf to the surface where its first contact with the air stimulates it to breathe. This

The great whales – the ultimate in Cetacean evolution. The Right (A) and Blue (B) whales are plankton feeders. They take enormous mouthfuls of sea water, straining planktonic shrimps and molluscs from the water as it pours through the curtain of fibrous baleen plates hanging from their upper jaws. The Sperm whale (C) has teeth on the bottom jaw and hunts for squid sometimes at great depths.

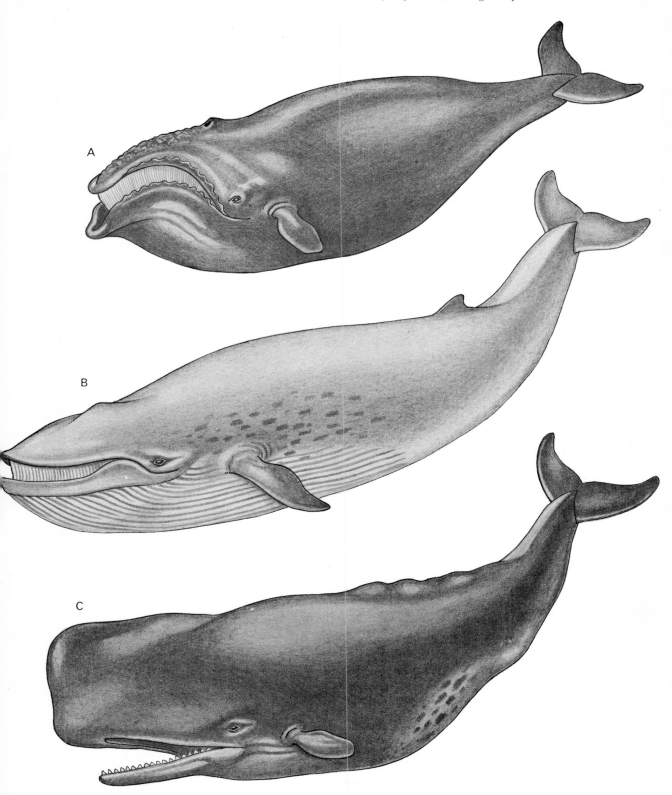

The under-surface of the upper jaw of one of the great filter-feeding whales. Planktonic crustaceans and the like are kept back by the fibrous baleen plates and are then dislodged by the tongue and swallowed.

Dolphins. Note the fish-like form, although they have horizontal tail-flukes, and only one pair of flippers which are equivalent to the fore-limbs.

powerful piece of parental behaviour may be responsible for records of dolphins and porpoises lifting to the surface people in danger from drowning. As air-breathing mammals they are particularly sensitive to the symptoms of respiratory distress; and will respond by aiding their offspring or companion to reach the air where they can breathe easily.

Soon after they are born, young whales are able to swim and breathe without help but, of course, they are, for up to a year, reliant upon their mother's milk, which is the richest in the world. It is thick and creamy, and contains about 50 per cent fat, 12 per cent protein, and 2 per cent sugar, and is injected forcefully by special muscles into the throat of the suckling whale.

Why whales?

When the sea was already dominated by fish, one may well wonder what kind of rôle these creatures could play, that could not be better performed by their cold-blooded vertebrate relations. The fact is that cetaceans are a group whose success in the sea has only recently been threatened by us with our fast boats and explosive harpoons, and even then only the larger whales have suffered.

We can only conjecture that they owed their edge in competitiveness with fish to the fact that they were mammals. Even before they entered the water, the Creodont ancestors of whales were doubtless very active mammals, with some form of social organisations and language and had fairly large brains. The whales carried these characteristics with them into the oceans where they gradually evolved into numerous kinds of communal fish hunters. Even today, whales and their kin are pack hunters with a remarkable degree of social co-ordination, facilitated by a language of clicks, whistles and moans. What is even more remarkable is the claim that dolphins have even been able to imitate our own voices – although transposed into a higher pitch – and that the learnt words

are used in the correct context. If this proves to be true, then this opens up the real possibility of another species being able to communicate with us on a meaningful level for the first time. The smaller cetaceans which have been used for experiments have comparatively large brains, with convoluted *cerebral cortexes* like our own. A bottle-nosed dolphin's brain weighs 3 pounds which represents 1.2 per cent of its body weight, and compares favourably with our own, which is also 3 pounds but is equivalent to 1.93 per cent of our total mass. Not surprisingly, these delightful mammals seem to display an awareness and 'intelligence' that is shown only by the higher Primates. Admittedly, it is difficult to construct 'intelligence' tests that are meaningful to cetaceans; they cannot manipulate with their flippers for instance; although they seem to make up for this inadequacy amply by using their mouths.

It may well be, then, that cetaceans were able to find a niche for themselves in the seas by outwitting fish like sharks for a share of their food. And because of their hot blood, they were able to exploit the planktonic resources of the cold seas girdling the Poles perhaps more effectively than fish. Here, increasing size became a decided advantage particularly for tapping the prodigious quantities of shrimps in the plankton. Many zoologists have expressed some surprise that 'life' could become organised into units as large as the 150-ton blue whales, particularly since these monsters exist by straining relatively miniscule food particles from the surface of the sea. There is, however, an interesting commercial message to be gleaned from considering the size of these animals, because they have grown big and prosperous by cutting out the middle men in the food chain.

8

Man, the presumptuous quadruped

by John Napier (with introduction by John Sparks)

Introduction

No survey of animal designs would be complete without considering ourselves. Although we have taken to a vertical way of life, balancing on our two hind legs, our bodies were originally designed to be supported and moved by four. In the process of becoming a vertical ape, a large number of modifications have had to be made to our chassis – some successful, others less so.

The most remarkable aspect of human walking is not that man walks so well but that he walks so badly. This somewhat provocative statement needs qualification. It should not be taken to mean that as individuals we are a gangling bunch of Johnnies-head-in-the-air who are liable to disappear down every unguarded manhole that looms in our path, or trip over every wrinkle in the carpet, but that – as a species – we are ill-equipped to carry out the complicated manoeuvres of bipedalism without a large proportion of our number paying the penalty in terms of injury or disablement.

Ovid saw human uprightness as a hallmark of man's questing spirit: 'God made man erect to contemplate the heavens.' All other things being equal, Ovid's is a somewhat suicidal precept which probably accounts for the number of individuals who actually do fall down open manholes.

The truth of the matter is that human walking is a very risky business that brings in its wake a remarkably long list of human ailments. The 'banana skins' of life await the unwary around every corner: the slick of oil, the uneven set of a paving-stone, the sudden push in the back are constant threats to our shaky equilibrium. During some phases of the walking cycle the fact that man does not fall flat on his face depends on the support of an area of foot not much bigger than a postage stamp.

With training it is possible to reduce the risks of bipedalism considerably. Karl Wallenda, leader of the famous German 'high-wire' family, The Great Wallendas, recently crossed the 700-foot-deep Tallulah Gorge in Georgia, USA, walking for nearly a quarter of a mile on a slack wire. Karl Wallenda even had the nerve and the energy to stop half-way across and stand on his head. Yet Sylvia Potts, a British athletic hope, slipped and fell, whilst in the lead and within a few feet of the finishing line, during the women's 400 metres event at the Edinburgh Commonwealth Games in 1970. Training is not the whole answer.

Man is not the only primate which puts itself at hazard by depending on two limbs. Gibbons which progress by swinging from overhead branches by two arms are notoriously accident prone. It has been reported that 33 per cent of a series of 200 specimens shot in Malaysia showed evidence of healed fractures presumably the result of locomotion accidents.

Theoretically, man would have been very much better off, as far as his physical well-being is concerned, if he had stuck to a four-legged gait. But even quadrupedalism has its own quota of disadvantages, but they are negative rather than positive. Where all four extremities are feet, suborned to the function of walking, there is little opportunity for manual skills to develop. In fact many people see human bipedalism as the first step which led to the emergence of human technology initiated by the freeing of the hands from the chores of locomotion. Tools can only be used or made by mammals when the hands are free. The sea otter, a well-known tool user, employs a rock to smash open abalones while floating on its back. Hooved animals don't use their hands at all for any activity other than support except perhaps for 'pawing' the ground. Carnivores, rodents and insectivores have paws – hands of a sort – and they do with them what they can but with no great success as they lack prehensility. Even among quadrupedal primates, which universally possess a degree of prehensility of the hand that permits them (for example) to feed using one hand only, have a restricted manual repertoire; their hands are essentially foot-hands. Only man has a true hand-hand.

Quadrupedalism has other disadvantages. By standing on four legs instead of two a quadruped effectively reduces its potential height by 50 per cent. This in itself is unimportant, but its effect on the range of vision is not. Many species of open-country mammals such as prairie marmots, mongooses, meerkats, bears and monkeys have evolved a bipedal stance as part of their normal behavioural repertoire. In this way their visual range can be doubled thus diminishing the danger of attacks by predators. Monkeys in particular rely heavily on visual surveillance as their sense of smell and hearing is not so highly developed as in other mammals.

The freeing of the hands from locomotor responsibilities and the augmentation of the visual range are two of the principal factors believed to be concerned in the development of a permanent form of bipedalism in man.

There is little room for doubt that, for all its disadvantages, the assumption of upright posture was the primary adaptation that led to the emergence of the human stock. All man's characteristics – his culture, his ability to speak and use a language, his technology and his large brain – might be looked upon as the consequences of standing and walking upright. Even if bipedalism is a mechanically imperfect adaptation the price we pay is, relatively speaking, a small one. After all, if the worst comes to the worst, we can still get by with inserts in our shoes, crutches and bath-chairs.

Evolution of the upright posture

There is good reason to believe that far from being unique to man, the upright posture is a primate heirloom that has been handed down from generation to generation for 50 million years. Fossil evidence from the geological time period known as the Eocene (starting some 55 million years ago and lasting for 18 million years) indicates quite clearly that at this time primates were small-bodied, long-legged creatures that clung to the boles of trees and to upright branches with their trunks held vertically and their legs acutely bent at hips and knees. Several of their descendents are found today among the large group of relatively primitive primates known as the prosimians (lemurs, galagos, tarsiers, etc.). Some of these early 'vertical clingers', as they are called, evolved into quadrupeds by the relatively simple adaptive process of shortening their hindlimbs and lengthening their forelimbs. One living lemur, the ring-tail ((Lemur catta) shows precisely the expected intermediate characters between vertical clinging and quadrupedalism.

It seems highly unlikely that the human stock arose directly from a vertical clinging ancestry, but there is good reason to believe that the ancient and basic possession of truncal uprightness has dominated primate postural evolution and blazed the trail that ultimately led to the emergence of man.

Bipedalism, surprisingly, is not unique to man. All higher primates (above the lemur grade of evolution) are capable of standing upright and a few actually walk upright. For example among the monkeys of the New World, the spider monkeys are great bipedalists. Even the relatively primitive squirrel monkey (Saimiri sciureus) given the appropriate stimulus becomes a bipedal walker. An experiment was carried out whereby the arms of an infant squirrel monkey were taped to its sides. Normally the infant would clutch its mother's fur with its hands but deprived of this resort it becomes quite helpless. The mother solves the problem by scooping up its infant in its arms and walking away on two legs.

The Old World monkeys rise to their hindlegs at the drop of a coconut so to speak. Indeed the need to transport food items from one place to another is a factor that commonly induces bipedalism in living monkeys. Carriage of food also gets chimpanzees up

on to their hindlegs. The Dutch zoologist, Adriaan Kortlandt, has filmed chimpanzees raiding native plantations in forest clearings and running back on two legs to the cover of the forest clutching as many paw-paws and sweet potatoes as they can carry. Gorillas, too, are frequent bipedalists. The males use the added height that a two-legged stance provides to make themselves more menacing when performing the threat display that culminates in their classic chest-beating demonstration.

Any or all of these activities that result in intermittent two-footedness are plausible behavioural precursors for the habitual bipedalism of man. In view of the ancientness and the universality of truncal uprightness in the primate order it is hardly surprising that natural selection should have operated so strongly and so rapidly in bringing about this transfiguration in our own family.

Pre-adaptation is a state of affairs in which characters already possessed by an animal suddenly come into their own, usually as a result of a major environmental change. Often they are behavioural characters; that is to say an animal species may show a tendency to perform such and such an activity without possessing the physiological or anatomical wherewithalls to perform it to the best advantage. If the particular behaviour becomes critical for some special aspect concerned with survival of the species, natural selection will operate rapidly and strongly on any genetic mutation or variation that crops up if, thereby, the performance of the activity is enhanced. In this situation with the established behavioural pattern forcing the pace, it is likely that the final product will be such that it is highly adaptive in one set of circumstances but wholly unsuitable in another. The idea that I am putting forward in this chapter is that human bipedalism is just such a case. The advantages in certain spheres of the human life-style are immense – in fact they couldn't have taken place without the upright posture – but the disadvantages from other points of view are equally great. When a quadrupedal animal like a monkey stands or walks on two legs the spine is curved and the hips are semi-flexed; in such a position the centre of gravity is displaced well in front of the area of support of the two feet, so that the animal is in a state of gross imbalance which can only result in its falling flat on its face. The reason that it doesn't do so is that its balance is restored by flexion of the knees which brings the centre of gravity once more over the feet. A curved back, bent hip and bent knee posture probably represent the base-line conditions from which man's ape-like ancestors started out on their pilgrimage to

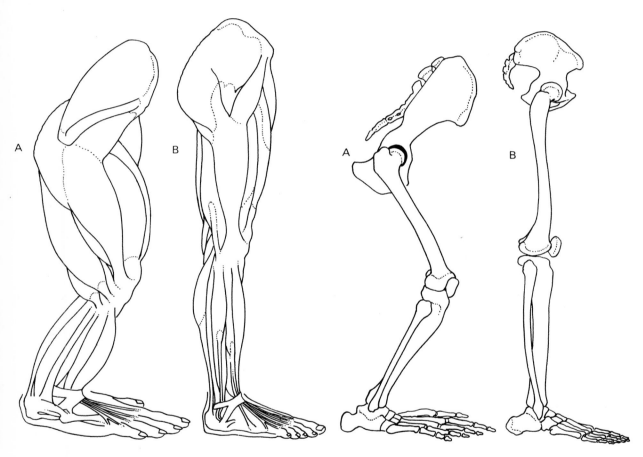

the primate Nirvana of upright bipedalism.

The anatomical changes that were necessary for this transition involved the straightening of the spinal column, the elongation of the hindlimbs, the shortening and broadening of the pelvis, adjustments to the muscles of the hip concerned with side-to-side balance during bipedal walking. As the legs developed the ability to straighten at the hips and to extend behind the body, the knees had to be straightened out too; this required an alteration in the relative proportions of the opposing muscle groups acting on the knee – the hamstrings (flexors)

and the quadriceps (extensors). In apes, the flexors are the dominant group while in man the extensors are more important.

Above all, the most critical changes to the upright posture have occurred in the foot. Principal among these was the development of an arched foot to replace a totally flat foot, and the realignment of the big toe whereby it lies parallel to the other toes instead of being splayed-out sideways as it is in all monkeys and apes.

This list is by no means comprehensive. No mention for instance has been made of the extensive modifica-

A young female baboon demonstrating her 'sitting pads'.

tions in the central nervous system (brain) underlying these changes in posture.

A matter that interests anthropologists, though in itself of little consequence, is the identification of the primary adaptation that initiated the whole bipedal bit. According to some authorities it was the development of the large buttock muscle, the *gluteus maximus* so powerful in man and so relatively insignificant in apes and monkeys, that was responsible. There is no doubt that man has massive buttocks compared with monkeys or chimpanzees who only have sitting pads on their bottoms. Human buttocks are not all muscle; there is a good deal of fat there too but the muscular element is what makes a human bottom a 'bum'. The *gluteus maximus*, for all its prominence, plays, rather surprisingly, a secondary rôle in man's ability to stand and walk on the level. The

gluteus maximus comes into its own however in climbing stairs or a steep slope where its chief function is to correct any tendency for the trunk to jack-knife on the legs as it is all set to do under the influence of gravity were the *gluteus maximus* not there to haul it back.

The most important muscles in the human hip region are the lesser glutei, technically called the *gluteus medius* and *gluteus minimus*. In man these muscles have shifted their pelvic attachment forwards so that their job is to stabilize the hip during the side-to-side rocking movements that occur when the weight is shifted from one side to the other during bipedal walking. Instability of the hip renders normal walking quite impossible as one sees only too clearly in a child with a congenital dislocation of the hip, a hip affected by tuberculosis or by osteochondritis (Perthé's Disease). Victims of these conditions waddle rather than walk.

In the monkey and ape the two muscles have much the same position and function as the *gluteus maximus* in man; they hold the trunk upright and this prevents the jack-knifing effect already referred to. In the transition from quadrupedal to bipedal posture these muscles assumed a new rôle as side-to-side stabilizers shifting their points of bony attachment accordingly. This change left both an actual as well as a functional gap which was filled by an enlargement of the already existing, but insignificant, *gluteus maximus*.

Diseases and injuries of the hip joint are common human complaints, the majority of which occur during the life periods of infancy, adolescence and old age. The childhood ailments have already been mentioned which leaves only the geriatric ones to be considered. Osteoarthritis is a common enough complaint that takes its toll from middle-age onwards and represents a mechanical breakdown of the joint cartilage of the hip causing stiffness and considerable pain. A fracture of the neck of the femur is the almost inevitable consequence of a bad fall in an

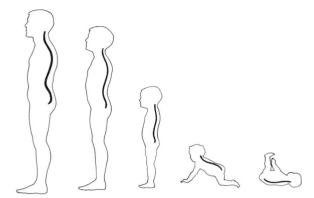

Changes in curvature of the human vertebral column from infancy to adulthood.

elderly person and is one of the commonest accidents in the book. Luckily the surgical treatment is simple and effective. However, it is no coincidence that hip surgery is one of the most highly developed branches of orthopaedic practice.

The curvature of the spinal column in most quadrupedal mammals forms a single span from skull to pelvis. In higher primates, such as the monkeys, the pitch of the arc is modified in the lumbar region where the curve is somewhat flattened. In the apes this tendency is further exaggerated, and in man it is converted into a distinct curve in the opposite direction. Changes in the human condition from birth to adulthood reflect, to some extent, the evolutionary story. The curve of the human infant's spinal column differs from the mammalian condition only marginally – by a reversal of curvature in the region of the neck. By the time the adult state has been reached the difference is more pronounced and an additional reversal – in the lumbar region – has taken place. The result is the development of a sinuous bony column having a forward curve in the neck region, a backward curve in the thoracic region, and another forward curve in the lumbar region. These curves are the mechanical answer to the biological problem of balancing the human body on a rather diminutive pair of feet.

The fact that the evolutionary progression from a quadrupedal to a bipedal type of spinal curvature is mimicked by the progression from the human infant to the human adult is not surprising when one recalls the dictum of the late Professor F. Wood Jones: 'The human child sits up before it stands; the human stock sat up before it stood'.

A further consequence of the upright posture is that humans have developed a larger and more backwardly curved sacrum.

Man's sacrum is larger and bulkier than in any ape or monkey which is reasonable when one appreciates the superincumbent weight it is required to carry. It is sharply bent backwards at its junction with the last lumbar vertebra. The angle thus formed between the plane of the sacrum and the plane of the spinal column measures 62° in man but only 32° in apes and even less in monkeys. This angle is increased in human females compared with males. This adaptation carries the seeds of its own destruction; it is a risky bit of biological engineering by which natural selection has attempted to steer a compromise between three conflicting needs: (1) the maintenance of the upright posture, (2) the preservation of a wide birth-canal, and (3) the mechanical stability of the lumbo-sacral articulation. In the event, natural selection has not been wholly successful inasmuch as this region is highly liable to mechanical failure. The load carried by the last lumbar vertebra may lead to a forward dislocation of this bone with dire consequences. Low back pain of various sorts including the ubiquitous 'lumbago', the prestigious slipped disc and the common or garden 'sciatica' are some of the hazards which man is heir to. It is my contention that the upright posture is a thoroughly risky business and this is nowhere so applicable as in the low back and pelvic regions.

We have hardly mentioned the hazards of birth that the new configuration of the human pelvis has brought in its wake. The female pelvis is as wide as it can be from side-to-side as well as fore-and-aft without interfering with the mechanical needs of the human walking gait. Any further widening would lead to an imperfect and physiologically uneconomic type of walk which, charitably, might be described as 'a rolling waddle'. The real nature of the problem becomes apparent when one appreciates that in human evolution not only has there been a trend towards uprightness, and therefore a reduction in the dimensions of the pelvic outlet, but also a trend towards enlargement of the brain at birth. While the infant's body constitutes only 5 per cent of its adult size, the brain is already 25 per cent of its final size. Thus there is a conflict between the needs for walking and the needs for childbearing. To me this suggests that natural selection in man, to coin a phrase, has got its priorities all wrong. However, there is a way out of this dilemma – thanks to resources of human culture, in the shape of improved postnatal care. The premature infant in western society is no longer at such risk as it was fifty years ago. One might anticipate that natural selection will now begin to redress the mistakes of the past and that a shortening of the human gestation period will eventually occur and infants will be born at a less mature state.

It is well known that modern women vary considerably in their pelvic dimensions. Without going too deeply into the matter there are the wide-hipped female-females and the narrow-hipped boyish-females. Sexual selection in civilised societies is tending towards the latter type rather than the former for reasons perhaps not unconnected with the fashionable trends in female emancipation which of course started 60–70 years ago. The soft and yielding Rubenesque female-female is *out* and the capable, slim-hipped Diana-the-Huntress is *in*.

The rôle of the sexes is changing in western culture, and nature (now the effect of culture rather than its cause) is following suit.

The sort of arguments adduced here to support the overall thesis that, in some situations, the upright posture provides great advantages but that in others it is a heavy burden, can be applied equally forcibly to the inguinal region (the site of hernias), to the knee joint, to the ankle and to the foot, but I think enough has been said already to make further examples unnecessary.

Man has gained much by evolving as a walking biped; he has gained, in fact, his uniqueness as a primate species, but he has to pay the price of the presumptuousness of his ancestors. He pays it when his feet are 'killing him', when his knees are 'not as good as they might be, doctor', when his back feels 'as if he had been burned with a red-hot poker', when he gets an aching, dragging pain 'down there'. Oh yes! make no mistake! Man is a rotten old biped. He would have been much better off if he had stuck to four legs – one at each corner. But then, of course, he wouldn't have been a man.

9
The Bird

Introduction

Defying gravity is not something to be lightly under-taken, as countless human 'jumpers' have found to their cost. Make-shift wings frantically flapped by the arms have never postponed the inevitable body shattering impact with the ground! Flight, no less than other means of moving around, necessitates good design – perhaps even more than usual; after all, if we stumble and trip over, the consequence may be no more than a grazed knee, but the results of fall-ing out of the sky could be fatal for all but the smallest creatures. The point is that flying animals have overcome more problems than less adventurous creatures, and in doing so can be credited with having taken living materials to their limits. In the problem of support, aquatic animals have few troubles; their chassis are relatively small and there seems little restriction on body bulk. Not so with living on the land, for immediately a sturdy skeleton becomes a necessity and being huge is not anywhere near so practical a proposition. Although walking on just two feet does require skill and tricky engineering, flying is exacting in all respects. The fuselage must be reinforced but, because of the tremendous energy needed for staying aloft, bulk imposes a severe power penalty so it must be kept to an absolute minimum. In the construction of a bird, therefore, it is possible to see how a lightweight air frame has been achieved without sacrificing strength.

Anyone who has carved up the Sunday chicken should be able to guess from the distribution of meat that birds are dual-purpose creatures. Although some species like swifts and albatrosses spend most of their lives in the air, none of the 8,580 known kinds have completely broken free of the need to visit the ground. Thus they have one locomotory and support system for flying. The other acts as a shock absorber, a built-in springboard as well as for effecting walking, climbing or swimming. So far as the overall design of a bird is concerned, the weight must be comfortably

supported at some time by the wings, and, at other times, on the legs without sacrificing stability both on the ground or in the air. These two totally indepen-dent means of transport have their origins in the history of these creatures.

Birds evolved from small lizard-like reptiles which were already bipedal – that is walking, mostly on their hind legs. They therefore already had a satis-factory arrangement for terrestrial locomotion, thus leaving the forelimbs or 'arms' free to evolve into wings without interfering with the operation of the legs. In the modern bird we find a body divided neatly into separate bone and muscular arrangements – one for flying, the other for walking.

Not all creatures have solved the problem of flying as elegantly as birds. Bats need their legs to keep the wing and tail membranes stretched, and, although their powers of flying are good, their mobility on the ground is hampered. Nevertheless, by hanging up-side down, they can make use of more space in caves than would be possible for an orthodox bird! Bats, however, can only beat birds at night, so it is possible to say that the avian design is the most versatile for flying ever to emerge from the vertebrate stable.

The airframe - or skeleton

Like the frames of our own aircraft, the bird's skeleton combines lightness with strength. A frigate bird weighing as much as a dressed chicken has only a 3- or 4-ounce skeleton, a remarkable achievement in weight reduction. Rigidity has been preserved by the fusion of many of the *vertebrae* or back bones – so that there are fewer independent bones in the avian body than in those of other vertebrates. The ribs also lock together by means of special backward projecting processes. Great strength is thereby achieved to withstand the stresses and strains of taking off, flying and landing. What mobility has been lost in the body itself is made up for by the highly articulated neck. Many of the long bones are also hollow, thus saving

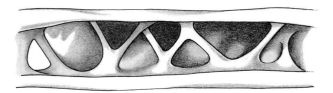

Cut-away part of a vulture's wing bone showing how the bone is formed into struts giving the structure immense strength without incurring too heavy a weight penalty.

on weight, but are reinforced by a lattice-work of fine struts which makes them incredibly strong. Light though they are, there is evidence that pterodactyls – extinct skin-winged reptiles – would have weighed much less than birds of equivalent size.

The skeleton of a flying bird is dominated by two extensive slabs of bone corresponding to the breast and shoulder, and the hips; these are known as the

pectoral and *pelvic* girdles respectively, and they serve to support the limbs and to provide anchorage points for the muscles which power them. Although these are not hefty, the thin bone is strengthened by ridges to prevent distortion while stressed by muscles working under load. The most interesting girdle is the pectoral, because this together with the muscles is the one which is concerned with flight.

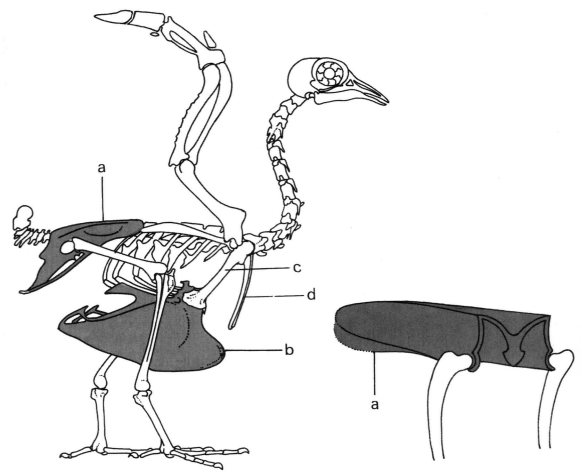

Skeleton of a bird showing the pelvic (a) and pectoral (b) girdles; the coracoid (c). The wishbone (d) helps to prevent the chest region from collapsing.

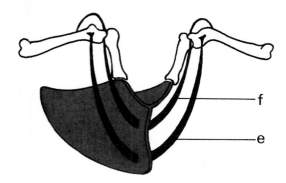

Diagrammatic transverse sections of pectoral girdle. (e) and (f) show the layout of pectoralis major *and the* supracoracoideus *which depresses and raises the wing respectively.*

The power plant

The muscles which move the wings account for up to one-third of the total body weight of a bird and these are mostly located on a keel-like extension of the breast. Most people will be acquainted with the flight muscles which form the pale succulent flesh on the breast of a well-roasted chicken, although in birds which are able to take to wing with more stamina than these domesticated jungle fowl this mass of muscle is dark red with *myoglobin*, a pigment that stores oxygen in readiness for the heavy demands of flight. In fact the breast flesh consists of two distinct muscles. The one that is carved up first, or the more superficial, is called the *pectoralis major*, and is inserted onto the *humerus*; it is by far the largest muscle, and when it contracts the wing is forced downwards on the power stroke. Deeper beneath the *pectoralis major* is a smaller one called the *supracoracoideus* which raises the wing, by virtue of the fact that the ligament from it runs upwards over the 'shoulder' and inserts onto the upper side of the humerus. Thus, the chief flight muscles are able to be located side by side on the sternum. Also they are neatly placed beneath the centre of lift to preserve the bird's balance during flight. So powerful are these muscles that the whole of the pectoral girdle is strengthened to prevent their combined pull from collapsing the fuselage. This danger is overcome by the development of a strong compression strut – the *coracoid* – which runs from the upper side of the keel practically to the shoulder. In the flying bird it is the 'wishbone's' function to brace the wings apart, and in strong flying species the bone takes on the form of a much shallower 'V' than in the earth-bound chicken.

Flying is a very energetic business, requiring a great deal of power, expended sometimes over long periods. A flying budgerigar, for instance, expends power up to twenty times faster than its resting rate – a performance which a person of extraordinary athletic ability would do well to emulate for a few minutes.

Also, weight for weight, the power developed by bird muscle is several times that of our own. One of the characteristics which enables them to summon forth such a sustained high output of energy is their high body temperature of at least 41°C (106°F). In their 'hotted up' bodies, all reactions can be accelerated and movements made faster. To maintain this high level of activity, a great deal of fuel and oxygen must be pumped round the body. Birds have comparatively large hearts; a sparrow's is 2.7 times larger than that of a mouse of comparable body weight. A humming bird's may account for 2.75 per cent of the body weight and beats at 1,000 times each minute. The respiratory tract of birds is unusual insofar as it is connected to a series of nine or so air sacs, and their development may be related to the bird's high demand for oxygen. Although it is dubious whether much oxygen passes through the membraneous walls of the sacs, the movement of air between them and the lungs may ensure that the highly vascular surfaces of the lungs are constantly being exposed to flowing air; in our own lungs, a great deal of the volume may contain stagnant air, so reducing the efficiency of the structure as a gas exchanger. With birds, no such problem may exist!

The air sacs may have yet another function. Like highly tuned engines, a flying bird may have a tendency to overheat, and although they can probably tolerate internal temperatures up to 110°F without harm, excessive heat is as lethal as excessive cold. Evaporation of water from the surface of the sacs may help to cool the breast muscles while they are in action.

How a bird flies

All flying machines need generously proportioned surfaces for producing lift, but these must be achieved without escalating weight. Bats and pterodactyls managed this by the development of highly elastic membranes which could be stretched between the

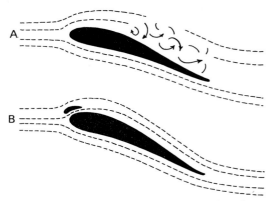

(A) When the angle of attack to the air flow increases too much for a given air speed, turbulence is created on the upper surface. The lift is destroyed and the wing is said to stall.

(B) The smooth or laminar air flow can be restored at high angles of attack by placing a slot on the leading edge of the wing.

fuselage and the arm and finger struts. Birds have feathers. These are made of *keratin*, the sort of substance that forms the dead outer layers of our skin. Each feather consists of a stiff central shaft and a blade, and the surface area is immense compared with the weight (thus the saying 'as light as a feather'). Apart from their streamlining and insulating functions, the most important part of the wing is composed of pinions that are specialised for beating against the air; they are elongated and especially stiff. The advantage of a wing composed of twenty or more primary and secondary pinion feathers is that the aerodynamic properties of the structure are not greatly affected by the loss of one or two. However, they are 'dead' structures and when badly damaged cannot be repaired in situ. Birds therefore have to moult regularly in order to maintain their aerial performance. Some wildfowl change all of their flight feathers at once, and during this period are grounded but most species renew them one at a time so as not to lose their ability to fly.

The aerodynamics of a bird are far more complicated than those of an aircraft. In the latter, power for forward movement is provided either by the thrust of a jet engine, or by one or more rapidly rotating propellers, and lift to support the machine in the air is produced by the wings. In birds, however, both the *lift* and *motive force* must be provided by the wings alone, and flapping is a means of generating as large a force as possible by increasing the air speed over them.

As a broad generalisation, it is possible to consider the inner part of the wing as being directly comparable to the wing of an aircraft because, in fast flight, it generates much of the lift; and the outer part, including the pinions, as being equivalent in function to the propeller because it provides a great deal of the propulsive force.

The inner wing

The whole of a bird's wing is shaped like a typical aerofoil. That is, it is rather thicker at the front, with a slightly convex upper surface. Air passing over a structure like this speeds up over the top surface, creating a reduction in pressure, so sucking the wing upwards. An upward thrust is thereby generated which manifests itself in the form of *lift*. The camber of the wing (or the curvature of the upper side), its angle of attack to the airstream, and the speed of the air will all effect the magnitude of the lift. Now, when a bird flies rapidly, the inner wing remains relatively still; it is held in the airstream at an angle which creates the most favourable degree of lift for the circumstances.

The outer wing

During rapid flight, this is the part of the wing which is flapped most, circumscribing a broadly elliptical or figure-of-eight course. From the raised position, it is brought downwards and, because of the pressure on the pinions, it is twisted forward. The lift forces generated by this part of the aerofoil now become deflected forwards because of the angle of inclination of the wing, and can thus be used to overcome the drag forces acting on the bird as it moves through the air. Lift generated by the outer wing can therefore be harnessed to power the bird.

During take off, when the air speed over the inner wing is too slow to generate much lift, the outer part is brought sharply forward to create sufficient lift to raise the bird off the ground. The propulsive force is also generated by the outer wing as it is flicked back-

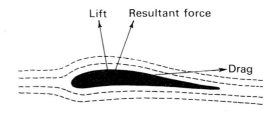

Lift   Resultant force

Drag

Gannet coming in to land with forewing slots raised (bastard wings) to prevent the wings from stalling at high angle of attack which is necessary to keep them working at slow air speeds.

wards on the recovery stroke. This, however, is a very tiring method of flying and can only be kept up for a few seconds. Heavy birds must taxi sometimes considerable distances before sufficient air speed is achieved to create enough lift to take them clear of the ground.

Stability

If the airflow over an aerofoil breaks into a series of eddies, and becomes *turbulent* as opposed to smooth or *laminar*, the wing no longer produces any lift, and is said to *stall*. This usually happens when the inclination of the wing to the airflow exceeds a critical angle. Birds have a number of devices which help to maintain *laminar* flow over their aerofoils at high angles of attack. A bunch of feathers attached to the remnant of the bird's 'thumb' aids boundary layer control and reduces the speed at which the wing stalls. Birds can

therefore keep their wings working at high angles of attack, and slow airspeeds, by extending their thumbs – or bastard wings. This is particularly important when they come in to land. Aircraft use precisely this method of preventing their wings stalling by bringing a series of fore-wing slots into operation when they take off and land. In fact, it was the investigation into bird flight that led to the application of slots to high speed aircraft wings. The slot principle also operates at the wing tips of soaring birds. These have emarginated pinion feathers, which, when the wing is fully extended at low air speeds, act like a series of slots to prevent the wing tips stalling.

Wings for different jobs

Birds use their ability to fly in different ways. It is not then surprising to find that those which need to

Perch

pursue fast moving insects have wing designs which differ from those of basically terrestrial species, which chiefly use their wings to propel them in an instant from a lurking enemy. Each bird has therefore evolved wings which allow it to survive best in its own particular way of life. By and large, however, wings tend to fall into four chief types.

The soaring wing

Vultures, buzzards, eagles, storks and pelicans among others have broad, highly cambered, wings which enable them to make every use of thermals. Their weight is spread out over a large surface area, and so the wing loading is comparatively low, giving a slow sinking speed in still air. Providing they find air rising faster than their sinking speed, they will gain height. Slotting both on the pinions and the fore-wing (thumb) helps to keep the air flow laminar and keeps the wings working at low speeds. This is particularly important at the wing tips because soarers have to be able to turn tightly in order to keep inside a thermal. Under these circumstances the 'inside' wing may be moving much slower than the outer one, and the slotted pinions therefore help to prevent the inner aerofoil from stalling. Control surfaces, such as the tail, are broad to promote manoeuvrability and increase lift at slow air speeds.

Nubian vulture, a thermal soarer. Note the deeply slotted pinions.

Bird landing on perch.

Perch

Oceanic gliders

Birds like albatrosses, man o' war birds, and fulmar petrels have long narrow wings, well adapted for operating in the high air speeds encountered over the sea. Their narrowness is useful in cutting down air resistance, but does mean that in comparison with soarers the weight of the bird is spread out over a smaller wing area. Accordingly, the wing loading is high and in calm air albatrosses find it difficult to generate sufficient lift to take off. Like the broad soaring wings, gliding wings are tiring to flap because of their size and inertia, and in still conditions birds like albatrosses are easily exhausted. Not surprisingly

then, albatrosses are associated with the oceanic trade winds, and were always recognised by sailors to be harbingers of gales. Moving in tight circles is not a necessary ability for these birds and so one finds little evidence of 'slotted' pinions in oceanic gliders.

It is a fascinating exercise to compare these birds with the larger pterodactyls, because these too were probably oceanic but were adapted to the much more gentle climate that prevailed at the end of the Cretaceous period. Cherry Bramwell and George Whitfield of Reading University have calculated that these light, large-winged reptiles could glide comfortably at 10–12 m.p.h., a speed at which most birds would fall out of the air.

The North Atlantic gannet, an oceanic glider. Note the absence of deeply slotted pinions.

A puffin (left) coming in to land. Notice the use of the tail and feet as air brakes. As the speed drops, the wing is tilted at a greater angle to the air stream to maintain lift. The bastard wing – or the forewing slot – is raised to keep the airflow smooth over the upper surface and prevents the wing from stalling. The feathers are beginning to ruffle near the shoulder which indicates that the wings are near to their stalling point.

The wings of the bottom bird are on the backward stroke and show how the pinion feathers part to enable the wing to recover with as little air resistance as possible.

(below) Tawny owl. These birds have low wing-loadings to enable them to fly slowly, and the pinions are covered with a velvet-like pile to cut down noise; the rodents upon which they prey are therefore more easily taken unawares.

Painted stork (left), about to touch down, with the undercarriage lowered to take the impact of landing. Note the broad wings with deeply slotted pinions adapted for soaring.

Wandering albatross (below), a high-speed oceanic soarer with long narrow wings that offer comparatively little resistance to the air. Note the absence of slotted pinions.

White pelicans in flight. These heavy birds have broad wings with slotted pinions which are adapted for soaring over the land. This helps them to journey from one lake to another.

High-speed wings

Swifts spend most of their lives chasing insects, and accordingly they have wings that can propel them comfortably at 60 m.p.h. – perhaps even 100 m.p.h. on occasions. Similarly, falcons that swoop onto their prey from great heights have wings that conform to the swift type; it is alleged that peregrines have reached 180 m.p.h. in a dive. A high-speed flying wing is narrow and swept back – to cut down air resistance. The tip is tapered partly to cut down the lift destroying effect of vortexes swirling round from the high-pressure areas beneath the wing to the upper surface. Slotting of the pinions would create too much drag and is hardly ever in evidence on such wings.

Wings for a quick launching

Many birds use their wings chiefly to get them out of trouble as quickly as possible. Since the bird itself may not be moving forward very quickly, the wings must be constructed so that they can be flapped extremely rapidly to create a lot of lift immediately they are called into action. Pheasants or partridges are good examples. They have broad but very short wings, which thus have low inertia, and high camber (very arched) to give powerful lift when the wings are whirred. The pinions are deeply slotted and twist forwards when moved sharply down through the air and provide a strong propulsive force. When disturbed, a heavy pheasant can rise steeply through the woodland canopy to glide to the safety of another spinney.

Other uses for wings

It is almost a general principle of life that structures that have evolved for one specific function will, sooner or later, become modified for other, quite different, purposes. Such is the case with the avian wing. The most obvious modification that springs to mind is that of the penguin's flipper, a tough, rigid structure which is beautifully adapted for flying in water. The other, perhaps less well-known fact, is that the wing can be used as a useful placard on which to place messages – or signals. The message may be quite simple conveying information to all and sundry about which species the signaller belongs to (such as the metallic patterns or specula on the wings of ducks) or rather more involved courtship signals which are designed to convey the mood of the signaller. Flying itself tends to make a noise, and in many species this sound has been used as the raw material for the development of non-vocal calls. The feathers of the wing can become modified to promote the effectiveness of the display. Thickened quills produce a snapping sound in Gould's manakin. The broad-tailed humming bird's leading pinion is notched and in flight makes a form of instrumental song used

Ruby throated humming bird showing the wings at the end of the forward stroke and completing the back stroke. This shows how the whole of the rather stiff wing is rotated at the shoulder to produce a helicopter effect. The wings are therefore able to provide lift on both the forward and backward strokes.

by the males to attract the females. In the case of the male Argus pheasant, the fine ocellated inner wing feathers are so long that flight is well nigh impossible. In this case, the display has taken complete precedence over the aerofoil function of the wing.

Limitations of size

It would take 100,000 bee humming birds to balance the weight of a 300-pound cock ostrich. This gives some idea of the range in size of modern birds. One thing is, however, quite obvious: ostriches and the other avian heavyweights have lost the power of flight somewhere during their evolutionary history. The reason for this is that muscle power increases more slowly than weight; and so as birds become larger and heavier they become *relatively* less powerful. A point is therefore reached when the development of greater

size for some or other advantage means giving up flying as a means of moving around, and taking permanently to the ground. Once this has taken place, weight is no problem and the advantages of being big can be exploited as has been the case with the rheas, emus, cassuwaries and ostriches. For vertebrates, the weight barrier to self-powered flight seems to be about 40 pounds, the weight of a cock kori bustard, which is the heaviest flying bird, and even this species spends much of its time on the ground.

Of course many other flying vertebrates have exceeded the kori bustard in dimensions, but it is dubious whether they exceeded it in weight. *Pteranodon* was a fantastic flying reptile that lived 60 million years ago and had membraneous wings spanning all of 26 feet. Nevertheless these 'dragons' had puny bodies with relatively little muscular development around the breasts and shoulders and weighed no more than 35 pounds. An extinct new world vulture, *Teratornis incredibilis* holds the avian record for wing span – 17 feet, but like its surviving relatives, the Andean and Californian condors with 10-foot wing spans, it was doubtless a soarer with a comparatively light body. The latter species just tips the scales at 20–25 pounds. A glider, the wandering albatross has the widest span today of 12 feet, but weighs only about 14 pounds.

If 40 pounds does represent the weight barrier for self-powered flight, then it does not hold up much hope for those who cherish the idea of us being able to fly under our own steam. Alas, even angels now seem to be a practical impossibility, requiring a four-foot keel, breast muscles to match, and hurricane force winds if they were to stand even the remotest chance of self levitation. At least we can console ourselves with the thought that fairies stood a sporting opportunity of being able to flit around the toadstools!

Some of the long distance travels undertaken by birds are so much more superior than those taken by earth-bound beasts that one is bound to ask the question whether this is due to anything inherent in

flight. Firstly, birds are able to sustain speeds of
30–50 m.p.h. for long periods without any trouble.
In any given time then, birds can travel much
further than mammals. Secondly, if the energy costs
of transporting a unit weight of animal over a given
distance are assessed, it turns out that flying is far less
energy consuming than walking or running. When
compared with a bird of the same weight, a mammal
uses ten to fifteen times more fuel to cover a given
distance, although, because of its slow speed, this
would be expended over a long period of time. It does,
however, become obvious why small mammals have
not evolved long seasonal migratory movements
which would place an almost impossibly high drain
on their resources. Small birds, using fat with nearly as
high an energy yield as petrol, burn up at least 1 per
cent of their body weight each hour of flight; this is a
figure that compares favourably with aircraft which
may consume anything between 2 and 36 per cent of
their weight in fuel per hour. Vance Tucker, a North
American scientist who has performed many elegant
experiments on the energetics of bird flight, has said
that a pigeon can fly more economically than a light
aircraft and that a Canada goose may be able to
perform better than a jet transport!

Of course birds move very fast and their rate of
using energy is correspondingly high – much higher
than our own. However, if one calculates the cost of
moving a unit weight of bird over a given distance,
then surprisingly, it turns out flying is a very cheap
kind of transport.

Index

Bibliography

ALEXANDER, R. MCN. *Animal mechanics*. Sidgwick & Jackson, hard and paperback, 1968.

ALEXANDER, R. MCN. *Size and shape*. E. Arnold, 1971.

BUCHSBAUM, R. *Animals without backbones*. 2 vols. Penguin Books, 1951.

CARRINGTON, R. *Elephants*. Chatto, 1958; Penguin Books, 1962, o.p.

CURRY, J. D. *Animal skeletons*. E. Arnold, hard and paperback, 1970.

GRAY, Sir J. *Animal locomotion* (World naturalist). Weidenfeld & Nicolson, 1968.

GRAY, Sir J. *How animals move*. C.U.P., 1953, o.p.

HARRISON, R. J. and KING, J. E. *Marine mammals*. Hutchinson, n.e. hard and paperback, 1968.

MARSHALL, N. B. *The life of fishes*. Weidenfeld & Nicolson, 1965.

MATTHEWS, L. H., ed. *The whale*. Allen & Unwin, 1969.

MORTON, J. E. *Molluscs*. Hutchinson, n.e. hard and paperback, 1967.

ROMER, A. S. *Man and the vertebrates*. 2 vols. Penguin Books, 1970.

THOMSON, Sir A. L., ed. *A new dictionary of birds*. Nelson, 1964.

TUCKER, V. A. *The energetics of bird flight*. Scientific American, May 1969, page 70.

WELLS, M. *Lower animals*. Weidenfeld & Nicolson, hard and paperback, 1968.

WHITFIELD, G. and BRAMWELL, C. *Palaeonengineering: birth of a new science*. New Scientist, vol. 52, no. 775, 1971, page 202.

YOUNG, J. Z. *The life of mammals*. O.U.P., 1957.